T0319430

Reference Frame Theory

IEEE Press Series on Power Engineering

Series Editor: M. E. El-Hawary, Dalhousie University, Halifax, Nova Scotia, Canada

The mission of IEEE Press Series on Power Engineering is to publish leading- edge books that cover the broad spectrum of current and forward-looking technologies in this fast-moving area. The series attracts highly acclaimed authors from industry/academia to provide accessible coverage of current and emerging topics in power engineering and allied fields. Our target audience includes the power engineering professional who is interested in enhancing their knowledge and perspective in their areas of interest.

Reference Frame Theory

Development and Applications

Paul C. Krause
PCKA

IEEE
PRESS
SERIES
ON POWER
ENGINEERING

IEEE PRESS

WILEY

Published by John Wiley & Sons, Inc., Hoboken, New Jersey.
Published simultaneously in Canada.

For general information on our other products and services or for technical support, please contact our Customer Care Department within the United States at (800) 762-2974, outside the United States at (317) 572-3993 or fax (317) 572-4002.

Wiley also publishes its books in a variety of electronic formats. Some content that appears in print may not be available in electronic formats. For more information about Wiley products, visit our web site at www.wiley.com.

Library of Congress Cataloging-in-Publication Data

Names: Krause, Paul C., author.
Title: Reference frame theory : development and applications / Paul C.
    Krause, PC Krause and Associates.
Description: First edition. | Hoboken, New Jersey : Wiley-IEEE Press,
    [2020] | Includes bibliographical reference and index.
Identifiers: LCCN 2020029315 (print) | LCCN 2020029316 (ebook) | ISBN
    9781119721635 (cloth) | ISBN 9781119721635 (paperback) | ISBN
    9781119721628 (adobe pdf) | ISBN 9781119721659 (epub)
Subjects: LCSH: Relativity (Physics)
Classification: LCC QC173.5 .K73 2020 (print) | LCC QC173.5 (ebook) | DDC
    530.11–dc23
LC record available at https://lccn.loc.gov/2020029315
LC ebook record available at https://lccn.loc.gov/2020029316

Set in 9.5/12.5pt STIXTwoText by SPi Global, Chennai, India

Contents

About the Author

Paul C. Krause received a BSEE in 1956, a BSME in 1957, and a MSEE in 1958 from the University of Nebraska. He received a PhD in EE in 1961 from Kansas University. He taught at several colleges for 52 years retiring in 2009, after 39 years as a full professor, with Purdue University. He is the 2010 recipient of the IEEE Nikola Tesla Award. He has published over 100 technical papers and four text books in Electric Machines and Drives with two in their third edition.

Preface

The change of variables was introduced to me at the beginning of the PhD program. At that time, it was a mystery as to where the transformation originated, and it remained a mystery for nearly 60 years. Although there has been considerable work in the area since Park wrote his 1929 paper, no one to my knowledge has set forth the basis of the transformation. With the advent of the computer and power electronics, reference frame theory has become necessary in the teaching of machines and drives. However, the concept of reference frame theory was difficult to teach since the transformation seemed to appear from out of the blue. One had to more or less accept that it worked.

In 2016, during the writing of "Introduction to Power and Drive Systems," the connection between Tesla's rotating magnetic field and reference frame theory was uncovered, and it then became clear that the change of variables made the substitute variables portray the correct view of Tesla's rotating magnetic field from a given reference frame. This gave meaning to the transformation, and it is now easier to understand. This information was later published in the third edition of "Electromechanical Motion Devices" and an IEEE Paper.

This book covers some of the aspects of reference frame theory that the author has been involved with during the past 60 years, from the arbitrary reference frame to the basis of reference frame theory and field orientation. Other work including neglecting stator transients, multiple reference frames, and the relation of the transformation to symmetrical components are also discussed.

It is interesting that reference frame theory is contained in Tesla's rotating magnetic field, which is the basis of all known real transformations. Moreover, since the arbitrary reference frame can be used to derive symmetrical components, Tesla's rotating magnetic field, although not as direct, can also be considered the basis for complex transformations such as the symmetrical component transformation. Thus, it appears that all transformations, real and complex, used in the power and drives areas can be traced back to Tesla's rotating magnetic field.

This book is written for the engineers working in the power and drives areas as a reference and for the graduate students who want to know more about the history and basis of reference frame theory. It can be used as a textbook, and some instructors may want their class to become more informed regarding reference frame theory and require it as a reference text.

Paul Krause

1

A Brief History of Reference Frame Theory

In the late 1920s, R. H. Park, a young MIT graduate working for GE, wrote a paper on a new method of analyzing a synchronous machine [1]. He formulated a change of variables that, in effect, replaced the variables (voltages, currents, and flux linkages) associated with the stator windings of a synchronous machine with variables associated with fictitious, sinusoidally distributed windings rotating at the electrical angular velocity of the rotor. This change of variables, which eliminated the position-varying inductances from the voltage equations, is often described as transforming or referring the stator variables to the rotor reference frame. Although he did not refer to this as "reference frame theory," it was its beginning.

This new approach to machine analysis found limited use until the advent of the computer. Perhaps a discussion written by C. H. Thomas in the late 1950s was the first to set forth a method of using Parks equations to establish a stable computer simulation of a synchronous machine which is still being used today [2]. This opened the door to the use of a change of variables to analyze problems involving electric machines since this same method of simulation is used today for the simulation of all synchronous and induction-type machines.

There have been numerous changes of variables that have been set forth after Park's work. In the late 1930s, C. H. Stanley [3] employed a change of variables in the analysis of induction machines. He showed that the rotor-position-dependent inductances in the voltage equations of an induction machine, which are due to electric circuits in relative motion, could be eliminated by replacing the rotor variables with substitute variables associated with sinusoidally distributed stationary windings. This is often described as transforming or referring the rotor variables to a frame of reference fixed in the stator or the stationary reference frame. About the same time, E. Clarke [4] set forth an algebraic transformation for three-phase stationary circuits to facilitate their steady-state and transient analyses of three-phase ac power systems. She referred to these substitute variables as alpha, beta, and zero components.

Reference Frame Theory: Development and Applications, First Edition. Paul C. Krause.
© 2021 John Wiley & Sons, Inc. Published 2021 by John Wiley & Sons, Inc.

In [5], G. Kron introduced a change of variables that eliminated the rotor-position-dependent inductances of a symmetrical induction machine by transforming both the stator and the rotor variables to a reference frame rotating in synchronism with the fundamental electrical angular velocity of the stator variables. This reference frame is commonly referred to as the synchronously rotating reference frame.

D. S. Brereton et al. [6] employed a change of variables that also eliminated the rotor-position-varying inductances of a symmetrical induction machine. This was accomplished by transforming the stator variables to a reference frame rotating at the electrical angular velocity of the rotor.

Park, Stanley, Kron, and Brereton et al. developed changes of variables each of which appeared to be unique. Consequently, each transformation was derived and treated separately in the literature until it was noted in 1965 [7] that all known real transformations used in machine analysis were contained in one transformation. The Arbitrary Reference Frame was introduced in [7] as a general reference frame that contained all known transformations simply by assigning the speed of the reference frame. For example, when ω, the speed of the q and d axes, is set equal to zero, we have Stanley's and Clarke's transformations; with $\omega = \omega_r$, we have Park's and Brereton's transformations; and when $\omega = \omega_e$, we have Kron's. Although this was an interesting observation, the connection to Tesla's rotating magnetic field was not made. Although it should have been, since moving from one reference frame to another changes only the frequency that we observe Tesla's rotating magnetic field.

In a recent paper [8], the connection between Tesla's rotating magnetic field and the arbitrary reference frame was set forth. It was shown that the transformation to the arbitrary reference frame was contained in Tesla's expression for the rotating magnetic field. Moreover, once the symmetrical stator and rotor are transformed to the arbitrary reference frame, we have the q and d voltage equations for all machines. The only thing that must be transformed are the flux-linkage equations for the machine being considered [9].

Up until the writing of [10], the transformations were given without any explanation as to the basis of the transformation. It was accepted without question. Although it was possible to obtain the transformation by referring the abc axes to a qd-axis, there was not an analytical basis for the transformation. This plagued machine analysts for nearly a hundred years.

During the writing of [10] it was found that the equation for Tesla's rotating magnetic field contained the basis we had all been trying to find since Park's work. This forms the machine analysis in [10] and was explained in [9]. This approach to machine analysis is the subject of the next two chapters. In Chapter 2, we refer Tesla's rotating magnetic field to a rotating axis. In Chapter 3, we establish the connection between Tesla's rotating magnetic field and reference frame theory.

References

1 Park, R.H. (1929). Two-reaction theory of synchronous machines – generalized method of analysis – part I. *AIEE Trans.* 48: 716–727.
2 Riaz, M. (1956). Analogue computer representations of synchronous generators in voltage-regulation studies. *Trans. AIEE Power App. Syst.* 75: 1178–1184. See discussion by C. H. Thomas.
3 Stanley, C.H. (1938). An analysis of the induction motor. *AIEE Trans.* 57 (Supplement): 751–755.
4 Clarke, E. (1943). *Circuit Analysis of A-C Power Systems, Vol. 1 – Symmetrical and Related Components.* New York: Wiley.
5 Kron, G. (1951). *Equivalent Circuits of Electric Machinery.* New York: Wiley.
6 Brereton, D.S., Lewis, D.G., and Young, C.G. (1957). Representation of induction motor loads during power system stability studies. *AIEE Trans.* 76: 451–461.
7 Krause, P.C. and Thomas, C.H. (1965). Simulation of symmetrical induction machinery. *IEEE Trans. Power App. Syst.* 84: 1038–1053.
8 Krause, P.C., Wasynczuk, O., O'Connell, T.C., and Hasan, M. (2018). Tesla's contribution to electric machine analysis. Presented at the 2018 Summer Meeting of IEEE (5-9 August 2018), Portland, OR.
9 Krause, P.C., Wasynczuk, O., Pekarek, S.D., and O'Connell, T.C. (2020). *Electromechanical Motion Devices*, 3e. Wiley, IEEE Press.
10 Krause, P.C., Wasynczuk, O., O'Connell, T.C., and Hasan, M. (2017). *Introduction to Power and Drive Systems.* New York: Wiley, IEEE Press.

References

2

Tesla's Rotating Magnetic Field

2.1   Introduction

It is told that when Tesla was a student he argued with his professor that there must be a better way to design an electric machine than the dc machine being demonstrated in class. Nearly a decade later, Tesla invented the induction machine where he employed the rotating magnetic field [1]. The rotating magnetic field has been referred to as the rotating air-gap mmf, rotating magnetic poles, and the rotating field. Not only is Tesla's rotating magnetic field the backbone of the ac machine operation, it is also the key to the analysis of ac machines. The importance of viewing Tesla's rotating magnetic field from any frame of reference cannot be overemphasized. We show that the transform to a reference frame establishes the substitute variables associated with the fictitious circuits to portray Tesla's rotating magnetic field as viewed from that reference frame [2, 3]. Understanding this connection is very helpful to machine and electric drive analysis. This chapter is devoted to establishing the concept of viewing Tesla's rotating magnetic field in the arbitrary reference frame.

2.2   Rotating Magnetic Field for Symmetrical Two-Phase Stator Windings

An idealized sinusoidally distributed winding is shown in Figure 2.2-1. Each circle, $\otimes$ or $\odot$, has nc_s coils. For the positive current i_{as} the winding distribution from $0 < \phi_s < 2\pi$ may be approximated as

$$N_{as} = N_p \sin \phi_s \quad \text{for } 0 < \phi_s < \pi \tag{2.2-1}$$

and from $\pi < \phi_s < 2\pi$ as

$$N_{as} = -N_p \sin \phi_s \quad \text{for } \pi < \phi_s < 2\pi \tag{2.2-2}$$

Reference Frame Theory: Development and Applications, First Edition. Paul C. Krause.
© 2021 John Wiley & Sons, Inc. Published 2021 by John Wiley & Sons, Inc.

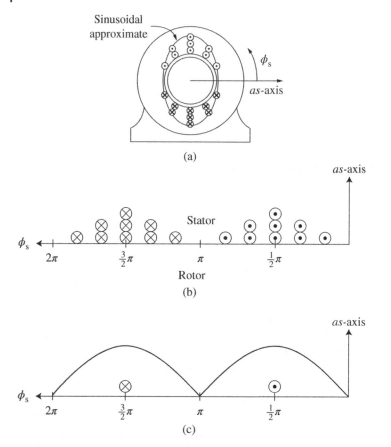

Figure 2.2-1 Sinusoidally distributed stator winding. (a) Approximate sinusoidal distribution. (b) Developed diagram of (a). (c) Sinusoidal approximation of (b).

where N_p is the peak turns density in turns/radian. If N_s represents the number of turns of the equivalent sinusoidally distributed winding, then, using ξ as a dummy variable of integration

$$N_s = \int_0^\pi N_p \sin \xi d\xi = 2N_p \qquad (2.2\text{-}3)$$

The developed diagram is the linear version of Figure 2.2-1a, and ϕ_s is positive from right to left. This diagram is valid since the air-gap length is very small compared to the radius of the rotor. Also, the $\otimes$ and $\odot$ are the notation that we use from here on for a sinusoidally distributed winding.

We can obtain a plot of the mmf by applying Amperes Law, $\oint \mathbf{H} \cdot d\mathbf{L} = i_n$, as shown in Figure 2.2-2, where one half of the mmf is dropped across each equal

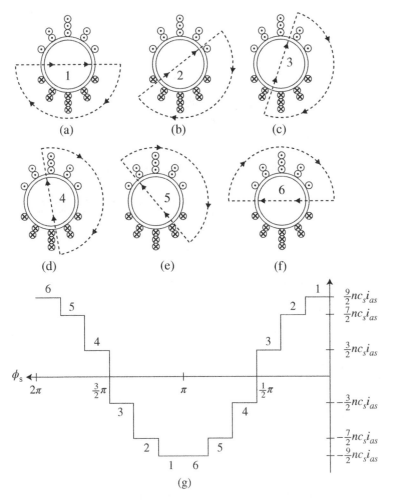

Figure 2.2-2 Closed paths of integration and the plot of mmf$_{as}$.

air gap. Figure 2.2-3 is a plot of the sinusoidal approximation of mmf$_{as}$ shown in Figure 2.2-2g.

A set of symmetrical, two-pole stator windings is shown in Figure 2.2-4. The windings are identical and sinusoidally distributed, and the air gap is uniform. Current enters the paper at ⊗ and out of the paper at ⊙. The rotor windings are not shown.

The mmf for each winding may be expressed as

$$\text{mmf}_{as} = \frac{N_s}{2} i_{as} \cos \phi_s \tag{2.2-4}$$

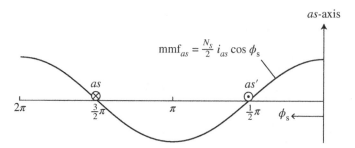

Figure 2.2-3 Plot of sinusoidal approximation of mmf$_{as}$ shown in Figure 2.2-2g.

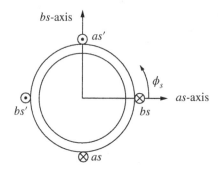

Figure 2.2-4 Elementary symmetrical, two-pole two-phase sinusoidally distributed stator windings.

$$\text{mmf}_{bs} = \frac{N_s}{2} i_{bs} \sin \phi_s \qquad (2.2\text{-}5)$$

The total air-gap mmf due to the two-pole stator windings, which is Tesla's rotating magnetic field would be

$$\text{mmf}_s = \text{mmf}_{as} + \text{mmf}_{bs}$$

$$= \frac{N_s}{2}(i_{as} \cos \phi_s + i_{bs} \sin \phi_s) \qquad (2.2\text{-}6)$$

This is a very important relationship. Let the steady-state currents be a balanced two-phase set

$$I_{as} = \sqrt{2}I_s \cos[\omega_e t + \theta_{esi}(0)] \qquad (2.2\text{-}7)$$

$$I_{bs} = \sqrt{2}I_s \sin[\omega_e t + \theta_{esi}(0)] \qquad (2.2\text{-}8)$$

where I_s is the rms value of the phase current, ω_e is the electrical angular velocity in rad/s, and $\theta_{esi}(0)$ is the phase angle of the currents. For this balanced set, $\tilde{I}_{bs} = -j\tilde{I}_{as}$.

If we substitute (2.2-7) and (2.2-8) into (2.2-6), we see that for this balanced set the rotating magnetic field travels counterclockwise at ω_e. Let us look at the rotating magnetic field from a rotating axis. To do this we introduce the q-axis, which rotates at an angular velocity ω. If we let θ be the angular position from the as-axis,

Figure 2.2-5 Symmetrical, two-pole two-phase sinusoidally distributed stator windings with a third magnetic axis (q-axis).

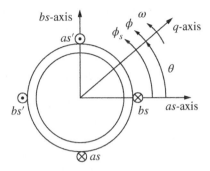

then

$$\theta = \omega t + \theta(0) \tag{2.2-9}$$

The displacement or coordinate ϕ shown in Figure 2.2-5 is from the q-axis just as ϕ_s is the displacement or coordinate from the as-axis. Let us relate this displacement to an adjacent displacement on the stator. From Figure 2.2-5

$$\phi_s = \theta + \phi \tag{2.2-10}$$

If we substitute (2.2-10) into (2.2-6), we obtain

$$\text{mmf}_s = \frac{N_s}{2}[i_{as}\cos(\theta + \phi) + i_{bs}\sin(\theta + \phi)] \tag{2.2-11}$$

The concept of Tesla's rotating field can be explained by assuming that the steady-state phase currents are (2.2-7) and (2.2-8) with $\theta_{esi}(0) = 0$. At time zero, I_{as} is maximum and I_{bs} is zero. The air-gap mmf is oriented along the as-axis. When the time has advanced to where $\omega_e t = \frac{\pi}{2}$, $I_{as} = 0$, and I_{bs} is maximum, the air-gap mmf has rotated from the as-axis to the bs-axis with an angular velocity of ω_e. That is, the mmf has rotated counterclockwise at ω_e relative to an observer on the stator.

Equation (2.2-11) is cornerstone to reference frame theory which we consider in Chapter 3; however, let us first deal with steady-state conditions. If we substitute balance steady-state currents given by (2.2-7) and (2.2-8) into (2.2-11) and after some work

$$\text{mmf}_s = \frac{N_s}{2}\sqrt{2}I_s\cos[(\omega_e - \omega)t + \theta_{esi}(0) - \phi] \tag{2.2-12}$$

Now, ω_e is the angular frequency of the electrical system, and ω is the angular velocity of the q-axis, where ϕ is the displacement from the q-axis, and $\theta(0)$ is the time zero position of the q-axis (2.2-9) which we have let to be zero in (2.2-12). We continue to set $\theta(0)$ equal to zero unless otherwise specified.

If now we let the angular velocity of the q-axis to be zero ($\omega = 0$) and $\theta(0) = 0$, then $\phi = \phi_s$, and we would be viewing Tesla's rotating magnetic field as a stationary observer. In this case, (2.2-12) becomes

$$\text{mmf}_s^s = \frac{N_s}{2}\sqrt{2}I_s \cos[\omega_e t + \theta_{esi}(0) - \phi_s] \qquad (2.2\text{-}13)$$

where we have added a superscript s to emphasize that we are observing Tesla's rotating magnetic field from a stationary frame of reference. If $\theta_{esi}(0)$ and ϕ_s are constants, mmf_s^s is viewed as a sinusoidal variation of frequency ω_e. If, for example, ϕ_s is zero, we would be positioned (fixed) at the as-axis ($\phi_s = 0$ in Figure 2.2-5), and the mmf_s^s would be pulsating at ω_e and the mmf is rotating counterclockwise.

Let us think about this for a minute. We are observing Tesla's rotating magnetic field as we stand on the stator. We see the rotating magnetic field pulsating at ω_e. It seems logical for this two-pole device that the variables associated with this rotating magnetic field in this frame of reference would vary at ω_e. This is something we already knew. That is, the balanced set of currents that produces this view of the rotating magnetic field is the balanced steady-state set given by (2.2-7) and (2.2-8). If we now let $\omega = \omega_e$, (2.2-12) becomes

$$\text{mmf}_s^e = \frac{N_s}{2}\sqrt{2}I_s \cos[\theta_{esi}(0) - \phi_e] \qquad (2.2\text{-}14)$$

where ϕ_e is the displacement from the q-axis which is rotating at ω_e. Since $\theta_{esi}(0)$ is constant for balanced steady-state conditions, (2.2-14) is a sinusoidal function of the spatial variable ϕ_e (the displacement from the q-axis). Its maximum value occurs at $\phi_e = \theta_{esi}(0)$, which is the phase angle of the balanced steady-state stator currents. In other words, we are observing Tesla's rotating magnetic field as we run at ω_e in the counterclockwise direction around the air gap, and this field appears constant to us for balanced steady-state conditions and the variables must be dc. Since ω^e is the angular frequency of I_{as} and I_{bs}, and the q-axis is also rotating at ω_e or in synchronism with the electrical system, this is referred to as the *synchronously rotating frame of reference*. Therein, since we have 360° vision, Tesla's rotating magnetic field appears to us as a sinusoidal function of ϕ_e (a space sinusoid) with the maximum value at $\phi_e = \theta_{esi}(0)$ from the q-axis and stationary relative to us since we are rotating at ω_e counterclockwise. This is depicted in Figure 2.2-6, where $\theta_{esi}(0)$ is selected negative to correspond to inductive circuits. A positive mmf_s^e is a south pole due to the stator currents (S^s) with the maximum intensity at $\phi_e = \theta_{esi}(0)$. A negative mmf_s^e represents a stator north pole (N^s).

We can view Tesla's rotating magnetic field from frames of reference other than a stationary observer. In Figure 2.2-6, the reference frame is rotating at ω_e in the counterclockwise direction. The importance of Figure 2.2-6 cannot be overemphasized. It is the connection between displacement and time. We find that this concept enables a clear visualization of the operation of electric machines and

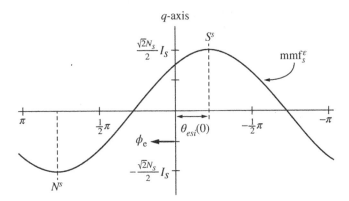

Figure 2.2-6 Tesla's balanced steady-state rotating magnetic field (mmf_s^e) viewed from $-\pi$ to π by an observer rotating counterclockwise about the air gap at ω_e with $\theta(0) = 0$ and $\theta_{esi}(0)$ negative. This is the synchronously rotating reference frame.

we use the synchronously rotating reference frame extensively in the analysis of machines and drives.

This is not the first time we have viewed a sinusoidal variation from a reference frame rotating at ω_e. When a phasor is rotated at ω_e in the counterclockwise direction, its real projection is the instantaneous value of the sinusoidal variation. When we consider the phasor as a constant we are running counterclockwise with it at ω_e or we have stopped rotation of the phasor. We relate variables in the synchronously rotating reference frame to the instantaneous and steady-state phasors.

2.3 Rotating Magnetic Field for Symmetrical Three-Phase Stator Windings

The arrangement of the stator windings of a two-pole three-phase device is shown in Figure 2.3-1. The windings are connected in a wye configuration (Y-connected), and they are identical, sinusoidally distributed with N_s equivalent turns, and with their magnetic axes displaced by $\frac{2}{3}\pi$. The positive direction of the magnetic axes is selected so as to achieve counterclockwise rotation of the rotating air-gap mmf with balanced stator currents of the abc sequence. The expressions for the air-gap mmfs established by the stator phases may be written by inspection of Figure 2.3-1. In particular,

$$\text{mmf}_{as} = \frac{N_s}{2} i_{as} \cos\phi_s \tag{2.3-1}$$

$$\text{mmf}_{bs} = \frac{N_s}{2} i_{bs} \cos\left(\phi_s - \frac{2}{3}\pi\right) \tag{2.3-2}$$

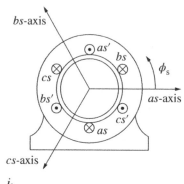

Figure 2.3-1 Elementary two-pole three-phase sinusoidally distributed stator windings.

$$\text{mmf}_{cs} = \frac{N_s}{2} i_{cs} \cos\left(\phi_s + \frac{2}{3}\pi\right)$$

(2.3-3)

As before, ϕ_s is the angular displacement about the stator measured from the as-axis. The total air-gap mmf_s is

$$\text{mmf}_s = \frac{N_s}{2}\left[i_{as}\cos\phi_s + i_{bs}\cos\left(\phi_s - \frac{2}{3}\pi\right) + i_{cs}\cos\left(\phi_s + \frac{2}{3}\pi\right)\right]$$

(2.3-4)

For balanced steady-state conditions, the stator currents for an *abc* sequence may be expressed as

$$I_{as} = \sqrt{2}I_s \cos[\omega_e t + \theta_{esi}(0)]$$

(2.3-5)

$$I_{bs} = \sqrt{2}I_s \cos\left[\omega_e t - \frac{2}{3}\pi + \theta_{esi}(0)\right]$$

(2.3-6)

$$I_{cs} = \sqrt{2}I_s \cos\left[\omega_e t + \frac{2}{3}\pi + \theta_{esi}(0)\right]$$

(2.3-7)

Substituting (2.2-10) for ϕ_s, and (2.3-5) through (2.3-7) into (2.3-4) and after some trigonometric manipulations we can obtain an expression for the rotating air-gap mmf (Tesla's rotating field) established by balanced steady-state stator currents as viewed from the *q*-axis.

$$\text{mmf}_s = \frac{N_s}{2}\sqrt{2}I_s\frac{3}{2}\cos[(\omega_e - \omega)t + \theta_{esi}(0) - \phi_s]$$

(2.3-8)

If mmf_s for the three-phase device given by (2.3-8) is compared with mmf_s for a two-phase device given by (2.2-12), we see that they are identical except that the amplitude for the three-phase device is $\frac{3}{2}$ times that of a two-phase device. It is convenient to first analyze the two-phase machine since the trigonometry is less involved, and once we have considered its "little sister," the extension to the three-phase machine is straightforward. It follows that we need not repeat the work we did regarding introducing the q-axis and the derivation leading up to (2.2-11). The only difference would be that (2.2-12) would be multiplied by $\frac{3}{2}$. Similarly, Figure 2.2-6 would apply to a three-phase stator if the amplitude of the magnetic field is multiplied by $\frac{3}{2}$.

2.4 Rotating Magnetic Field for Symmetrical Two-Phase Rotor Windings

In order to produce a constant torque, the rotor mmf, mmf_r, must rotate in unison with the stator mmf, mmf_s. Although we deal with torque in detail in later chapters, it is appropriate to take a first look at the means of producing torque in a symmetrical machine from the standpoint of stator and rotor mmfs.

The derivation of the air-gap mmf of symmetrical rotating circuits parallels that of the stationary circuits. As shown in Figure 2.4-1, the ar and br windings are orthogonal, and if they are identical and sinusoidally distributed, Tesla's rotating magnetic field of these symmetrical windings may be expressed.

$$\mathrm{mmf}_r = \frac{N_r}{2}(i_{ar}\cos\phi_r + i_{br}\sin\phi_r) \tag{2.4-1}$$

where N_r is the equivalent number of turns, and ϕ_r is the angular displacement from the ar-axis which is positioned by θ_r from the as-axis. In Figure 2.4-1, the positions θ_r and θ are referenced to the as-axis. Although we find that rotor windings are generally not sinusoidally distributed, the net rotor mmf is commonly approximated as a sinusoidal function of space and time as the stator mmf.

Figure 2.4-1 Two-phase rotating, identical, and sinusoidally distributed symmetrical windings.

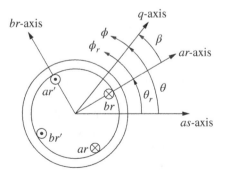

Now, the positions are related as

$$\phi_r = \beta + \phi \tag{2.4-2}$$

Substituting (2.4-2) into mmf$_r$, which is similar in form to (2.2-11), that is

$$
\begin{aligned}
\text{mmf}_r &= \frac{N_r}{2}[i_{ar}\cos\phi_r + i_{br}\sin\phi_r] \\
&= \frac{N_r}{2}[i_{ar}\cos(\beta + \phi) + i_{br}\sin(\beta + \phi)]
\end{aligned}
\tag{2.4-3}
$$

We will return to (2.4-3) later; however, let us continue our focus on the rotating air-gap mmf during balanced steady-state operation.

Now, from Figure 2.4-1

$$\beta = \theta - \theta_r \tag{2.4-4}$$

and during steady-state operation

$$\beta = (\omega - \omega_r)t + \theta(0) - \theta_r(0) \tag{2.4-5}$$

We find that during balanced steady-state operation of an induction motor, the rotor windings are short-circuited, and the rotor currents, which are induced by mmf$_s$, are balanced with the angular frequency of $(\omega_e - \omega_r)$. Thus, for a symmetrical machine let the currents be

$$I_{ar} = \sqrt{2}I_r\cos[(\omega_e - \omega_r)t + \theta_{eri}(0)] \tag{2.4-6}$$

$$I_{br} = \sqrt{2}I_r\sin[(\omega_e - \omega_r)t + \theta_{eri}(0)] \tag{2.4-7}$$

If we now substitute (2.4-5), (2.4-6), and (2.4-7) into (2.4-3), we obtain an expression for mmf$_r$

$$\text{mmf}_r = \frac{N_r}{2}\sqrt{2}I_r\cos\left[(\omega_e - \omega_r - \omega + \omega_r)t + \theta_{eri}(0) - \theta(0) + \theta_r(0) - \phi\right] \tag{2.4-8}$$

Now, we select $\theta(0)$ and $\theta_r(0)$ to be zero unless otherwise specified, whereupon (2.4-8) may be written as

$$\text{mmf}_r = \frac{N_r}{2}\sqrt{2}I_r\cos[(\omega_e - \omega)t + \theta_{eri}(0) - \phi] \tag{2.4-9}$$

Equation (2.4-9) is similar in form to mmf$_s$ given by (2.2-12). This is a very important observation. If the rotor currents are balanced, and if the angular frequency of these currents is $(\omega_e - \omega_r)$, which we assumed to derive (2.4-9), then the stator mmf$_s$ and the rotor mmf$_r$ will travel around the air gap at the same angular velocity. If, for example, we observe mmf$_s$ and mmf$_r$ as a stationary observer $(\omega = 0)$, mmf$_s^s$ and mmf$_r^s$ would both be traveling counterclockwise at ω_e relative to us. If we are riding on the rotor $(\omega = \omega_r)$, mmf$_s^r$ and mmf$_r^r$ would be traveling at $(\omega_e - \omega_r)$ counterclockwise relative to us. If we are running at ω_e, mmf_s^e and mmf_r^e would appear to us as constant space sinusoids of ϕ_e. The value of each will depend on ϕ_e. Since mmf_s and mmf_r are traveling in unison, a constant torque (power) is produced. We are starting to uncover Tesla's contribution to electric power generation and utilization.

Synchronously rotating reference frame; $\omega = \omega_e$ and $\phi = \phi_e$

$$\text{mmf}_s^e = \frac{N_s}{2}\sqrt{2}I_s \cos[\theta_{esi}(0) - \phi_e] \tag{2.4-10}$$

$$\text{mmf}_r^e = \frac{N_r}{2}\sqrt{2}I_r \cos[\theta_{eri}(0) - \phi_e] \tag{2.4-11}$$

We find that (2.4-10) and (2.4-11) are key to the animation and visualization of induction machine operation. Equations (2.4-10) and (2.4-11) are for two-phase machines. For three-phase machines, multiply each by $\frac{3}{2}$. The magnitude of the constant torque is determined by the number of turns of stator or rotor windings, the magnitude of the stator and rotor currents, and the relative positions of mmf_s and mmf_r. The relative position of the phase between the mmfs, $\theta_{esi}(0)$ and $\theta_{eri}(0)$, is easily discerned in the synchronous reference frame.

In the dc machine, torque is produced by the interaction of two stationary, orthogonal mmfs. A synchronous machine develops a constant torque only at synchronous speed, $\omega_r = \omega_e$. This occurs due to the fact that the rotor field which is stationary with respect to the rotor is rotating in unison with Tesla's rotating magnetic field. The induction machine produces torque at any rotor speed except when $\omega_r = \omega_e$. It is at this rotor speed that current is not induced in the short-circuited rotor circuits, and mmf_r disappears and mmf_s has no other mmf with which to interact. Please do not misinterpret (2.4-10) and (2.4-11); the only restriction on (2.4-11) is that $\omega_r \neq \omega_e$. The induction machine is operating at some rotor speed other than ω_e, and we are observing mmf_s and mmf_r from the synchronous reference frame. Regardless of from which reference frame we observe these mmfs, steady-state mmf_s and mmf_r are stationary relative to each other.

2.5 Rotating Magnetic Field for Symmetrical Three-Phase Rotor Windings

For the purpose of completeness we consider the three-phase rotor, even though we know the outcome. Figure 2.5-1 shows the three-phase rotor windings. We are going to assume that the windings may be actual windings rather than a squirrel cage rotor; however, the squirrel cage may be approximated by the windings shown in Figure 2.5-1.

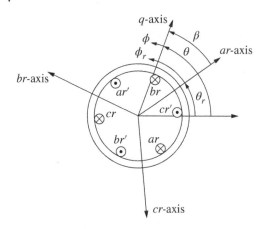

Figure 2.5-1 Three-phase rotating, identical, and sinusoidally distributed symmetrical windings.

The mmf$_r$ may be expressed as

$$\text{mmf}_r = \frac{N_r}{2} \left[i_{ar} \cos \phi_r + i_{br} \cos \left(\phi_r - \frac{2\pi}{3} \right) + i_{cr} \cos \left(\phi_r + \frac{2\pi}{3} \right) \right] \tag{2.5-1}$$

Now positions are related as

$$\phi_r = \beta + \phi \tag{2.5-2}$$

Substituting (2.5-2) into (2.5-1) yields

$$\text{mmf}_r = \frac{N_r}{2} \left[i_{ar} \cos(\beta + \phi) + i_{br} \cos \left(\beta + \phi - \frac{2\pi}{3} \right) + i_{cr} \cos \left(\beta + \phi + \frac{2\pi}{3} \right) \right] \tag{2.5-3}$$

Now, from Figure 2.5-1

$$\beta = \theta - \theta_r \tag{2.5-4}$$

In the steady state

$$\beta = (\omega - \omega_r)\, t + \theta(0) - \theta_r(0) \tag{2.5-5}$$

and

$$I_{ar} = \sqrt{2}\, I_r \cos[(\omega_e - \omega_r)t + \theta_{eri}(0)] \tag{2.5-6}$$

$$I_{br} = \sqrt{2}\, I_r \cos \left[(\omega_e - \omega_r)t - \frac{2\pi}{3} + \theta_{eri}(0) \right] \tag{2.5-7}$$

$$I_{cr} = \sqrt{2}\, I_r \cos \left[(\omega_e - \omega_r)t + \frac{2\pi}{3} + \theta_{eri}(0) \right] \tag{2.5-8}$$

Substituting (2.5-5) and (2.5-6) through (2.5-8) into (2.5-3), the rotor speeds cancel, and using trigonometric identities, we have, with $\theta(0)$ and $\theta_r(0)$ set to zero

$$\text{mmf}_r = \frac{N_r}{2} \sqrt{2}\, I_r \frac{3}{2} \cos[(\omega_e - \omega)\, t + \theta_{eri}(0) - \phi] \tag{2.5-9}$$

which is very similar to (2.2-12). If we let $\omega = \omega_e$, then

$$\text{mmf}_r^e = \frac{N_r}{2}\sqrt{2}\,I_r\,\frac{3}{2}\,\cos[\theta_{eri}(0) - \phi_e] \tag{2.5-10}$$

which is Figure 2.2-6, with the stator quantities changed to rotor quantities and the amplitude multiplied by $\frac{3}{2}$.

We see that (2.3-8) with $\omega = \omega_e$ is (2.5-10) and (2.5-9) are, respectively, (2.4-10) and (2.4-11) multiplied by $\frac{3}{2}$. Therefore, the comments following (2.4-10) and (2.4-11) apply to three-phase machine.

2.6 Closing Comments

We have established the rotating magnetic field for stator and rotor windings. We found that if the rotor currents in the short-circuited windings are a balanced set of frequency $(\omega_e - \omega_r)$, the rotating magnetic field established by the stator and rotor rotates in unison except when $\omega_r = \omega_e$ and no current is introduced in the rotor windings.

It is interesting that for $\omega < \omega_e$, the mmfs travel from right to left or counterclockwise toward synchronous speed; however, when $\omega > \omega_e$ they travel from left to right or clockwise toward synchronous speed. We will talk more about this in the next chapter.

References

1 Tesla, N. (1888). Electromagnetic motor. US Patent 0,381,968, 1 May 1888.
2 Krause, P.C., Wasynczuk, O., O'Connell, T.C., and Hasan, M. (2018). Tesla's contribution to electric machine analysis. Presented at the 2018 Summer Meeting of IEEE (5–9 August 2018), Portland, OR.
3 Krause, P.C., Wasynczuk, O., Pekarek, S.D., and O'Connell, T.C. (2020). *Electromechanical Motion Devices*, 3e. New York: Wiley, IEEE Press.

3

Tesla's Rotating Magnetic Field and Reference Frame Theory

3.1 Introduction

As the applications for converter-controlled electric drives began to emerge, it became evident that the traditional steady-state approach to electric machine analysis and the machine characteristics as portrayed by this type of analysis were inadequate to analyze and understand most electric drive systems. In particular, rapid electronic switching enabled control techniques which altered the performance of the electric machine beyond what one could envision from the traditional steady-state analysis. It became obvious that the transient characteristics of the electric machine had to be considered and that computer simulations of the dynamic performance of converter-controlled electric drives were useful, if not necessary, in the design of these drives. It was shown that these needs could only be addressed through the appropriate application of reference frame theory. Unfortunately, reference frame theory was considered more or less an abstract approach to electric machine analysis. This made it difficult to convey the concept of reference frame theory. Fortunately, the connection between Tesla's rotating magnetic field and reference frame theory provides a direct and straightforward analytical basis for viewing the physical variables of symmetrical windings from any frame of reference [1, 2].

Reference frame theory is nothing more than establishing the change of variables between the physical and substitute variables associated with the fictitious circuits that produce the rotating magnetic field as viewed from a given reference frame. Fortunately, the transformation that establishes the substitute variables (voltage equations and flux linkage equations) associated with the fictitious windings is determined directly from Tesla's rotating magnetic field. Reference frame theory is a straightforward extension of Tesla's rotating magnetic field and essential to the derivation of the machine equations and the fundamentals of machine control in electric drive systems.

Reference Frame Theory: Development and Applications, First Edition. Paul C. Krause.
© 2021 John Wiley & Sons, Inc. Published 2021 by John Wiley & Sons, Inc.

3.2 Transformation of Two-Phase Symmetrical Stator Variables to the Arbitrary Reference Frame

For the purpose of convenience, we repeat some equations from the previous chapter. Tesla's rotating magnetic field for the two-pole two-phase symmetrical, stator windings axes shown in Figure 3.2-1 is expressed by (2.2-6) as

$$\text{mmf}_s = \frac{N_s}{2}(i_{as}\cos\phi_s + i_{bs}\sin\phi_s) \tag{3.2-1}$$

From (2.2-10), the displacement coordinates ϕ and ϕ_s are related as

$$\phi_s = \theta + \phi \tag{3.2-2}$$

where θ is the position of the q-axis, ϕ is the displacement from the q-axis, and ϕ_s is the displacement about the stator periphery measured from the as-axis. We have also added a ds winding and axis. We will talk about this in more detail in a moment.

Substituting (3.2-2) into (3.2-1) yields (2.2-11), that is

$$\text{mmf}_s = \frac{N_s}{2}[i_{as}\cos(\theta + \phi) + i_{bs}\sin(\theta + \phi)] \tag{3.2-3}$$

In Chapter 2, we substituted steady-state currents, I_{as} and I_{bs}, into (2.2-11) and viewed the mmf$_s$ from the q-axis which could rotate at any angular velocity or remain stationary. The purpose now is to determine the change of variables that will portray Tesla's rotating magnetic field from any frame of reference, the *arbitrary reference frame*. To do this we first express (3.2-3) as

$$\text{mmf}_s = \frac{N_s}{2}\cos\phi(i_{as}\cos\theta + i_{bs}\sin\theta) + \frac{N_s}{2}\sin\phi(-i_{as}\sin\theta + i_{bs}\cos\theta) \tag{3.2-4}$$

Since mmf is in ampere-turns, the terms in the parentheses are currents; then, $\frac{N_s}{2}\cos\phi$ and $\frac{N_s}{2}\sin\phi$ can be interpreted as the result of fictitious orthogonal

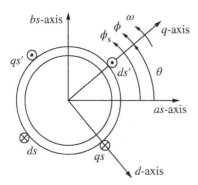

Figure 3.2-1 Two-pole two-phase sinusoidally distributed fictitious windings.

sinusoidally distributed windings. From Chapter 2 we realize that the winding distribution is orthogonal to the mmf distribution. When ϕ is zero, $\frac{N_s}{2}\cos\phi$ is symmetrical about the q-axis, and the current i_{qs} flowing in this q-axis fictitious winding is labeled qs in Figure 3.2-1. The current i_{qs} is

$$i_{qs} = i_{as}\cos\theta + i_{bs}\sin\theta \qquad (3.2\text{-}5)$$

When $\phi = \pm\frac{\pi}{2}$, the second fictitious winding, labeled ds in Fig 3.2-1, is symmetrical about an axis orthogonal to the q-axis. We call this axis the d-axis, as indicated in Figure 3.2-1, and we will select a positive d-axis at $\phi = -\frac{\pi}{2}$. From (3.2-4), we see that the current flowing in this winding is

$$i_{ds} = i_{as}\sin\theta - i_{bs}\cos\theta \qquad (3.2\text{-}6)$$

We can now write (3.2-4) as

$$\text{mmf}_s = \frac{N_s}{2}(i_{qs}\cos\phi - i_{ds}\sin\phi) \qquad (3.2\text{-}7)$$

Equation (3.2-1) is an expression for Tesla's rotating magnetic field in terms of physical windings and variables. The actual currents i_{as} and i_{bs} are flowing in sinusoidally distributed real-life stationary windings. On the other hand, the currents i_{qs} and i_{ds} in (3.2-7) are arbitrary-reference-frame substitute currents flowing in fictitious windings. These substitute variables are related to the physical variables, i.e. the substitute currents are the physical currents as viewed from the arbitrary reference frame. The fictitious windings are rotating at an arbitrary angular velocity of ω and are orthogonal and sinusoidally distributed with an identical number of equivalent turns and resistance as the actual windings. The substitute currents, i_{qs} and i_{ds}, that are flowing in these fictitious windings are related to i_{as} and i_{bs} by (3.2-5) and (3.2-6), respectively. We can think of the qs and ds windings as fictitious windings moving "through" the stator iron with i_{qs} and i_{ds} flowing in them, respectively.

The as and bs currents are transformed by the change of variables given by (3.2-5) and (3.2-6) to i_{qs} and i_{ds}, respectively. Since the voltages, currents, and flux linkages are functionally related, the same change of variables applies to all. Thus, the transformation to the arbitrary reference frame for the voltages, currents, and flux linkages may be written in matrix form as

$$\begin{bmatrix} f_{qs} \\ f_{ds} \end{bmatrix} = \begin{bmatrix} \cos\theta & \sin\theta \\ \sin\theta & -\cos\theta \end{bmatrix} \begin{bmatrix} f_{as} \\ f_{bs} \end{bmatrix} \qquad (3.2\text{-}8)$$

Symbolically

$$\mathbf{f}_{qds} = \mathbf{K}_s\mathbf{f}_{abs} \qquad (3.2\text{-}9)$$

where

$$(\mathbf{f}_{qds})^T = [f_{qs} \quad f_{ds}] \qquad (3.2\text{-}10)$$

$$(\mathbf{f}_{abs})^T = [f_{as} \quad f_{bs}] \tag{3.2-11}$$

$$\mathbf{K}_s = \begin{bmatrix} \cos\theta & \sin\theta \\ \sin\theta & -\cos\theta \end{bmatrix} \qquad \cdot \ (3.2\text{-}12)$$

and T denotes the transpose of a matrix. The angular position of the arbitrary reference frame is

$$\frac{d\theta}{dt} = \omega(t) \tag{3.2-13}$$

The substitute variables are the variables associated with the fictitious circuits that produce Tesla's rotating mmf as viewed from a given frame of reference. The subscript s is used in qs, ds, and $\mathbf{K}_s$ to distinguish stator physical variables from the variables associated with the rotor circuits. Later, we designate rotor variables with an r in the subscript. Finally, the angular velocity and angular position are related by (3.2-13). It is interesting that for this choice of the q- and d-axis locations, $\mathbf{K}_s$ is equal to $(\mathbf{K}_s)^{-1}$.

Note that we have not assigned a value to ω or θ. The angular position θ must be continuous; however, the angular velocity associated with the change of variables is unspecified. That is, the frame of reference from which we wish to view Tesla's rotating magnetic field may rotate at any constant or varying angular velocity or it may remain stationary. In other words, the angular velocity of the transformation is unspecified (arbitrary) and can be selected to specify the frame of reference that will expedite the solution of the system equations or to satisfy system constraints. Although we have used Tesla's rotating magnetic field to derive the change of variables, the transformation may be applied to variables of any waveform and time sequence and may have no relation to an electric machine. In an ab-sequence, the a-phase variables lead the b-phase variables by $\frac{\pi}{2}$ for balanced steady-state conditions; however, as we just mentioned, this does not imply that the variables must be a balanced steady-state set nor must the set be associated with a machine.

Let us assume that

$$v_{as} = r_s i_{as} + p\lambda_{as} \tag{3.2-14}$$

$$v_{bs} = r_s i_{bs} + p\lambda_{bs} \tag{3.2-15}$$

In matrix form

$$\mathbf{v}_{abs} = \mathbf{r}_s \mathbf{i}_{abs} + p\boldsymbol{\lambda}_{abs} \tag{3.2-16}$$

Now, to transform to the arbitrary reference frame

$$\mathbf{v}_{qds} = \mathbf{K}_s \mathbf{v}_{abs}$$
$$= \mathbf{K}_s \mathbf{r}_s (\mathbf{K}_s)^{-1} \mathbf{i}_{qds} + \mathbf{K}_s p (\mathbf{K}_s)^{-1} \boldsymbol{\lambda}_{qds} \tag{3.2-17}$$

Working on the first term on the right-hand side of (3.2-17)

$$\mathbf{K}_s \mathbf{r}_s (\mathbf{K}_s)^{-1} \mathbf{i}_{qds} = \mathbf{K}_s \begin{bmatrix} r_s & 0 \\ 0 & r_s \end{bmatrix} (\mathbf{K}_s)^{-1} \begin{bmatrix} i_{qs} \\ i_{ds} \end{bmatrix} = \mathbf{r}_s \mathbf{i}_{qds} \tag{3.2-18}$$

The second term on the right-hand side of (3.2-17) can be written as

$$\mathbf{K}_s p (\mathbf{K}_s)^{-1} \lambda_{qds} = \mathbf{K}_s p \left[(\mathbf{K}_s)^{-1} \right] \lambda_{qds} + \mathbf{K}_s (\mathbf{K}_s)^{-1} p \lambda_{qds} \tag{3.2-19}$$

The last term on the right-hand side of (3.2-19) is $p\lambda_{qds}$, and the first term becomes

$$\mathbf{K}_s p [(\mathbf{K}_s)^{-1}] \lambda_{qds} = \begin{bmatrix} \cos\theta & \sin\theta \\ \sin\theta & -\cos\theta \end{bmatrix} \omega \begin{bmatrix} -\sin\theta & \cos\theta \\ \cos\theta & \sin\theta \end{bmatrix} \lambda_{qds}$$

$$= \omega \begin{bmatrix} 0 & 1 \\ -1 & 0 \end{bmatrix} \begin{bmatrix} \lambda_{qs} \\ \lambda_{ds} \end{bmatrix} = \omega \begin{bmatrix} \lambda_{ds} \\ -\lambda_{qs} \end{bmatrix} \tag{3.2-20}$$

In expanded form

$$v_{qs} = r_s i_{qs} + \omega \lambda_{ds} + p \lambda_{qs} \tag{3.2-21}$$

$$v_{ds} = r_s i_{ds} - \omega \lambda_{qs} + p \lambda_{ds} \tag{3.2-22}$$

We have set forth the transformation to the arbitrary reference frame for variables associated with the stator circuits. It is important to understand that the change of variables is nothing more than a transformation of the stator variables to a selected frame of reference so that the substitute variables flowing in the fictitious reference frame windings provide the correct view of Tesla's rotating magnetic field from that frame of reference. In a later section, we consider rotor circuits and find that the transformation for symmetrical rotating circuits is very similar. In fact, we could have set forth just one transformation to a reference frame rotating at an arbitrary angular velocity which could be used for either stationary or rotating symmetrical circuits. That is, the transformation for stationary circuits can then be obtained by fixing the angular velocity of rotating circuits at zero. Although this approach is valid, it tends to be confusing; therefore, the transformation of stationary and rotor circuits are treated separately.

It is interesting that when we substituted $\theta + \phi$ for ϕ_s into (3.2-1) we leave the real world and entered the world of Tesla's rotating magnetic field where the variables are no longer as and bs but qs and ds for compactness. The analysis becomes based on the rotating magnetic field and we must think in terms of rotating with it and yet at any time, we can refer our analyses back to the real world. Once one becomes familiar with this concept, reference frame theory becomes straightforward.

The instantaneous power of a two-phase system is

$$P = v_{as} i_{as} + v_{bs} i_{bs} \tag{3.2-23}$$

In substitute variables, (3.2-23) becomes

$$P = (v_{qs} \cos \theta + v_{ds} \sin \theta)(i_{qs} \cos \theta + i_{ds} \sin \theta)$$
$$+ (v_{qs} \sin \theta - v_{ds} \cos \theta)(i_{qs} \sin \theta - i_{ds} \cos \theta) \qquad (3.2\text{-}24)$$

Equation (3.2-24) reduces to

$$P = v_{qs}i_{qs} + v_{ds}i_{ds} \qquad (3.2\text{-}25)$$

Now, (3.2-25) is in arbitrary-reference-frame variables, and it is equal to (3.2-23). Therefore, we have just shown that the power of a two-phase system is the same in all reference frames. Although the frequency of the voltage and current is reference frame dependent, since these variables must portray the view of Tesla's rotating magnetic field from a specific frame of reference, the average power is not a function of frequency and therefore not altered by the frame of reference from which we view Tesla's rotating magnetic field.

3.3 Transformation of Two-Phase Symmetrical Rotor Variables to the Arbitrary Reference Frame

Tesla's rotating magnetic field for rotating symmetrical sinusoidally distributed windings was set forth in Chapter 2. We repeat some of the figures and equations for convenience. Figure 3.3-1 is a repetition of Figure 2.4-1. An expression of Tesla's rotating magnetic field due to the rotating circuits is (2.4-1), which is

$$\text{mmf}_r = \frac{N_r}{2}(i_{ar} \cos \phi_r + i_{br} \sin \phi_r) \qquad (3.3\text{-}1)$$

In (3.3-1), N_r is the equivalent number of turns, and ϕ_r is the angular displacement from the ar-axis which is displaced by θ_r from the as-axis. In Figure 3.3-1,

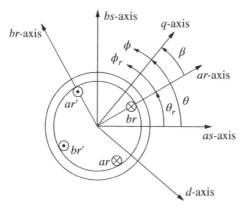

Figure 3.3-1 Two-phase rotating, identical, and sinusoidally distributed windings.

θ_r and θ are referenced to the as-axis. Note that the q-axis in Figure 3.3-1 is the same q-axis as in Figure 3.2-1. The displacements are related as

$$\phi_r = \beta + \phi \tag{3.3-2}$$

where

$$\beta = \theta - \theta_r \tag{3.3-3}$$

and θ and ϕ are the same as in Figure 3.2-1.

Substituting (3.3-2) into (3.3-1) yields

$$\text{mmf}_r = \frac{N_r}{2}[i_{ar}\cos(\beta + \phi) + i_{br}\sin(\beta + \phi)] \tag{3.3-4}$$

which is (2.4-3). If $\theta_r(0)$ and $\theta(0)$ are selected to be zero, and if the rotor windings are short circuited whereupon the frequency of the rotor currents is $\omega_e - \omega_r$, then from (2.4-9), mmf_r^e, with $\omega = \omega_e$, may be written as

$$\text{mmf}_r^e = \frac{N_r}{2}\sqrt{2}I_r\cos[\theta_{eri}(0) - \phi_e] \tag{3.3-5}$$

An observer running counterclockwise at ω_e with 360° vision would see a stationary mmf_r^e, which would be similar to mmf_s^e shown in Figure 2.2-6, differing by the turns, current, and the current phase angle.

The transformation for the rotor variables to the arbitrary reference frame follows that set forth in Section 3.2 for the stationary circuits. In particular, (3.3-4) may be written as

$$\text{mmf}_r = \frac{N_r}{2}\cos\phi(i_{ar}\cos\beta + i_{br}\sin\beta) + \frac{N_r}{2}\sin\phi(-i_{ar}\sin\beta + i_{br}\cos\beta) \tag{3.3-6}$$

If we let $\phi = 0$, we obtain i_{qr}, and if $\phi = -\frac{\pi}{2}$, we obtain i_{dr}. Equation (3.3-6) may now be written as

$$\text{mmf}_r = \frac{N_r}{2}(i_{qr}\cos\phi - i_{dr}\sin\phi) \tag{3.3-7}$$

The fictitious qr and dr windings for the rotor are shown in Figure 3.3-2. It is important to mention that if a device has symmetrical stator windings and symmetrical rotor windings, as is the case of an induction machine, the stator and rotor windings have symmetrical fictitious windings in the arbitrary reference frame. Moreover, in the arbitrary reference frame the fictitious windings are all stationary, relative to each other. This is significant since we find that the stator and rotor position-dependent mutual inductances are eliminated in all reference frames in the case of the symmetrical machine. Conversely, we find that in the case of the synchronous machine, the θ_r dependency of the mutual inductances is eliminated only in the rotor reference frame due to the magnetic asymmetry of a salient-pole machine as well as the fact that the rotor is not equipped with symmetrical windings.

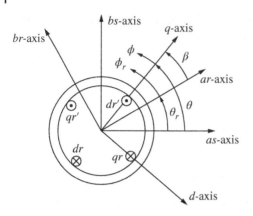

Figure 3.3-2 The fictitious *qr* and *dr* windings.

The transformation for the rotating symmetrical circuits becomes

$$\begin{bmatrix} f_{qr} \\ f_{dr} \end{bmatrix} = \begin{bmatrix} \cos\beta & \sin\beta \\ \sin\beta & -\cos\beta \end{bmatrix} \begin{bmatrix} f_{ar} \\ f_{br} \end{bmatrix} \tag{3.3-8}$$

or

$$\mathbf{f}_{qdr} = \mathbf{K}_r \mathbf{f}_{abr} \tag{3.3-9}$$

and

$$(\mathbf{K}_r)^{-1} = \mathbf{K}_r \tag{3.3-10}$$

The substitute voltage equations for a resistive-inductive circuit become

$$v_{qr} = r_r i_{qr} + (\omega - \omega_r)\lambda_{dr} + p\lambda_{qr} \tag{3.3-11}$$

$$v_{dr} = r_r i_{dr} - (\omega - \omega_r)\lambda_{qr} + p\lambda_{dr} \tag{3.3-12}$$

These substitute voltage equations are very similar in form to those for stationary circuits given by (3.2-21) and (3.2-22). In particular, β replaces θ in the transformation, and $\omega - \omega_r$ replaces ω in the voltage equations.

Before proceeding, it is interesting to note that if $\theta_r = 0$ in (3.3-3), it becomes the transformation to the arbitrary reference frame for the stationary circuits. Therefore, (3.3-8) is actually the only transformation that is needed for stationary and rotating symmetrical circuits.

3.4 Transformation of Three-Phase Stator and Rotor Variables to the Arbitrary Reference Frame

The three-phase stator is shown in Figure 3.4-1, which is a repetition of Figure 2.3-1 with the *q* and *d* axes added. Also, a repetition of (2.3-4) is (3.4-1):

$$\text{mmf}_s = \frac{N_s}{2}\left[i_{as}\cos\phi_s + i_{bs}\cos\left(\phi_s - \frac{2}{3}\pi\right) + i_{cs}\cos\left(\phi_s + \frac{2}{3}\pi\right)\right] \tag{3.4-1}$$

Figure 3.4-1 Elementary two-pole three-phase sinusoidally distributed stator windings.

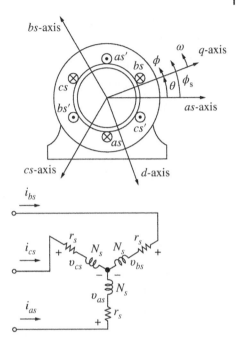

Now

$$\phi_s = \theta + \phi \tag{3.4-2}$$

which is a repetition of (3.2-2). Substituting (3.4-2) into (3.4-1) yields

$$\text{mmf}_s = \frac{N_s}{2}\left[i_{as}\cos(\theta+\phi) + i_{bs}\cos\left(\theta+\phi-\frac{2}{3}\pi\right) + i_{cs}\cos\left(\theta+\phi+\frac{2}{3}\pi\right)\right] \tag{3.4-3}$$

which after some work can be expressed as

$$\text{mmf}_s = \frac{N_s}{2}\cos\phi\left[i_{as}\cos\theta + i_{bs}\cos\left(\theta-\frac{2}{3}\pi\right) + i_{cs}\cos\left(\theta+\frac{2}{3}\pi\right)\right]$$
$$- \frac{N_s}{2}\sin\phi\left[i_{as}\sin\theta + i_{bs}\sin\left(\theta-\frac{2}{3}\pi\right) + i_{cs}\sin\left(\theta+\frac{2}{3}\pi\right)\right] \tag{3.4-4}$$

If we let $\phi = 0$, we obtain an expression for mmf$_s$ along the q-axis, and if $\phi = -\pi/2$, we obtain mmf$_s$ along the d-axis. Thus, the terms inside "[ ]" are i_{qs}, and the terms inside "[ ]" on the second line are i_{ds}. This forms the basis for the transformation of stator variables in three-phase machines. Because we are transforming three variables we need a third variable in addition to the q and d variables. It is referred to as the *zero variable*, which is independent of θ and is denoted as f_{0s}, where $f_{0s} = \frac{1}{3}(f_{as} + f_{bs} + f_{cs})$.

The commonly used three-phase transformation for the stator variables is given by Park [3] as

$$\mathbf{f}_{qd0s} = \mathbf{K}_s \mathbf{f}_{abcs} \tag{3.4-5}$$

where, for an *abc*-sequence

$$(\mathbf{f}_{qd0s})^T = [f_{qs} \quad f_{ds} \quad f_{0s}] \tag{3.4-6}$$

$$(\mathbf{f}_{abcs})^T = [f_{as} \quad f_{bs} \quad f_{cs}] \tag{3.4-7}$$

$$\mathbf{K}_s = \frac{2}{3} \begin{bmatrix} \cos\theta & \cos\left(\theta - \frac{2}{3}\pi\right) & \cos\left(\theta + \frac{2}{3}\pi\right) \\ \sin\theta & \sin\left(\theta - \frac{2}{3}\pi\right) & \sin\left(\theta + \frac{2}{3}\pi\right) \\ \frac{1}{2} & \frac{1}{2} & \frac{1}{2} \end{bmatrix} \tag{3.4-8}$$

$$\frac{d\theta}{dt} = \omega(t) \tag{3.4-9}$$

It can be shown that for the inverse transformation, we have

$$(\mathbf{K}_s)^{-1} = \begin{bmatrix} \cos\theta & \sin\theta & 1 \\ \cos\left(\theta - \frac{2}{3}\pi\right) & \sin\left(\theta - \frac{2}{3}\pi\right) & 1 \\ \cos\left(\theta + \frac{2}{3}\pi\right) & \sin\left(\theta + \frac{2}{3}\pi\right) & 1 \end{bmatrix} \tag{3.4-10}$$

Park chose to use a $\frac{2}{3}$ factor in $\mathbf{K}_s$. Tesla's rotating magnetic field (3.4-4) does not suggest the $\frac{2}{3}$ factor. Although it is unclear why Park used the $\frac{2}{3}$ factor, it is more or less irrelevant. We only need to remember that when evaluating power and torque using Park's variables, we must multiply by $\frac{3}{2}$ to compensate for the $\frac{2}{3}$ factor in the three-phase transformation.

Perhaps the best way to justify this is if we calculate power for a single-phase system, it is $v_{as}i_{as}$, which becomes $V_s I_s \cos\phi_{pf}$; for a two-phase system, it is $v_{as}i_{as} + v_{bs}i_{cs}$ or $v_{qs}i_{qs} + v_{ds}i_{ds}$, which becomes $2V_s I_s \cos\phi_{pf}$. Now since Park's $\frac{2}{3}$ factor makes $v_{qs}i_{qs} + v_{ds}i_{ds}$, the three-phase power is $2V_s I_s \cos\phi_{pf}$; when it should be $3V_s I_s \cos\phi_{pf}$, we must multiply by $\frac{3}{2}$.

Let us assume that

$$v_{as} = r_s i_{as} + p\lambda_{as} \tag{3.4-11}$$

$$v_{bs} = r_s i_{bs} + p\lambda_{bs} \tag{3.4-12}$$

$$v_{cs} = r_s i_{cs} + p\lambda_{cs} \tag{3.4-13}$$

If (3.4-8) through (3.4-10) are transformed to the arbitrary reference frame, we obtain

$$v_{qs} = r_s i_{qs} + \omega\lambda_{ds} + p\lambda_{qs} \tag{3.4-14}$$

$$v_{ds} = r_s i_{ds} - \omega \lambda_{qs} + p\lambda_{ds} \tag{3.4-15}$$

$$v_{0s} = r_s i_{0s} + p\lambda_{0s} \tag{3.4-16}$$

Although the $0s$ voltage equation has been added, the v_{qs} and v_{ds} equations are identical in form to those for two-phase systems. This is a result of the $\frac{2}{3}$ factor.

The $0s$ variables are zero for a balanced three-phase system. Also, three-phase electric machines are often connected in wye (Y) without a neutral conductor. In this case, the phase currents sum to zero. Since i_{0s} is zero, and if the system is symmetrical, v_{0s} becomes zero.

Let us consider a balanced, abc-sequence, three-phase set of variables

$$f_{as} = \sqrt{2} f_s \cos \theta_{esf} \tag{3.4-17}$$

$$f_{bs} = \sqrt{2} f_s \cos \left(\theta_{esf} - \frac{2}{3}\pi \right) \tag{3.4-18}$$

$$f_{cs} = \sqrt{2} f_s \cos \left(\theta_{esf} + \frac{2}{3}\pi \right) \tag{3.4-19}$$

Using (3.4-2) and trigonometric identities to transform to the arbitrary reference frame yields

$$f_{qs} = \sqrt{2} f_s \cos(\theta_{esf} - \theta) \tag{3.4-20}$$

$$f_{ds} = -\sqrt{2} f_s \sin(\theta_{esf} - \theta) \tag{3.4-21}$$

$$f_{0s} = 0 \tag{3.4-22}$$

Please note that if f_{0s} is zero, the three-phase system becomes a two-phase system. Therefore, the material in Section 3.3 also applies to a balanced three-phase system, except when calculating torque and power in terms of q and d variable, we must multiply by $\frac{3}{2}$ to make up for the $\frac{2}{3}$ factor in (3.4-8).

The three-phase rotor is shown in Figure 3.4-2, which is a repetition of Figure 2.5-1. We have added the d-axis.

The rotor mmf is given by (2.5-1), which is repeated here

$$\text{mmf}_r = \frac{N_r}{2} \left[i_{ar} \cos \phi_r + i_{br} \cos \left(\phi_r - \frac{2\pi}{3} \right) + i_{cr} \cos \left(\phi_r + \frac{2\pi}{3} \right) \right] \tag{3.4-23}$$

Now

$$\phi_r = \beta + \phi \tag{3.4-24}$$

where

$$\beta = \theta - \theta_r \tag{3.4-25}$$

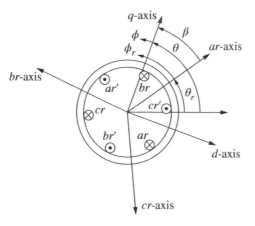

Figure 3.4-2 Elementary two-pole three-phase sinusoidally distributed rotor windings.

Substituting (3.4-24) into (3.4-23) yields

$$\text{mmf}_r = \frac{N_r}{2}\left[i_{ar}\cos(\beta+\phi) + i_{br}\cos\left(\beta+\phi-\frac{2\pi}{3}\right) + i_{cr}\cos\left(\beta+\phi+\frac{2\pi}{3}\right)\right]$$

(3.4-26)

After some work, (3.4-26) may be written as

$$\text{mmf}_r = \frac{N_r}{2}\cos\phi\left[i_{ar}\cos\beta + i_{br}\cos\left(\beta-\frac{2}{3}\pi\right) + i_{cr}\cos\left(\beta+\frac{2}{3}\pi\right)\right]$$
$$-\frac{N_r}{3}\sin\phi\left[i_{ar}\sin\beta + i_{br}\sin\left(\beta-\frac{2}{3}\pi\right) + i_{cr}\sin\left(\beta+\frac{2}{3}\pi\right)\right]$$

(3.4-27)

Here, i_{qr} is the terms within the first [ ], and i_{dr} is the terms within the second [ ]. The rotor zero quantity is $f_{0r} = \frac{1}{3}(f_{ar}+f_{br}+f_{cr})$. Park's transformation for the rotor variables becomes

$$\mathbf{f}_{qd0r} = \mathbf{K}_r\mathbf{f}_{abcr}$$

(3.4-28)

where f_{0r} is an abc-sequence

$$(\mathbf{f}_{qd0r})^T = [f_{qr}\, f_{dr}\, f_{0r}]$$

(3.4-29)

$$(\mathbf{f}_{abcr})^T = [f_{ar}\, f_{br}\, f_{cr}]$$

(3.4-30)

$$\mathbf{K}_r = \frac{2}{3}\begin{bmatrix} \cos\beta & \cos\left(\beta-\frac{2}{3}\pi\right) & \cos\left(\beta+\frac{2}{3}\pi\right) \\ \sin\beta & \sin\left(\beta-\frac{2}{3}\pi\right) & \sin\left(\beta+\frac{2}{3}\pi\right) \\ \frac{1}{2} & \frac{1}{2} & \frac{1}{2} \end{bmatrix}$$

(3.4-31)

$$(\mathbf{K}_r)^{-1} = \begin{bmatrix} \cos\beta & \sin\beta & 1 \\ \cos\left(\beta - \frac{2}{3}\pi\right) & \sin\left(\beta - \frac{2}{3}\pi\right) & 1 \\ \cos\left(\beta + \frac{2}{3}\pi\right) & \sin\left(\beta + \frac{2}{3}\pi\right) & 1 \end{bmatrix} \tag{3.4-32}$$

where

$$\frac{d\beta}{dt} = \omega(t) - \omega_r(t) \tag{3.4-33}$$

Let us assume that

$$v_{ar} = r_r i_{ar} + p\lambda_{ar} \tag{3.4-34}$$

$$v_{br} = r_r i_{br} + p\lambda_{br} \tag{3.4-35}$$

$$v_{cr} = r_r i_{cr} + p\lambda_{cr} \tag{3.4-36}$$

If (3.4-34) through (3.4-36) are transformed to the arbitrary reference frame, we have for the balanced conditions

$$v_{qr} = r_r i_{qr} + (\omega - \omega_r)\lambda_{dr} + p\lambda_{qr} \tag{3.4-37}$$

$$v_{dr} = r_r i_{dr} - (\omega - \omega_r)\lambda_{qr} + p\lambda_{dr} \tag{3.4-38}$$

$$v_{0r} = 0 \tag{3.4-39}$$

3.5 Balanced Steady-State Stator Variables Viewed from Any Reference Frame

A balanced two-phase set may be expressed as

$$f_{as} = \sqrt{2} f_s \cos\theta_{esf} \tag{3.5-1}$$

$$f_{bs} = \sqrt{2} f_s \sin\theta_{esf} \tag{3.5-2}$$

The above equations form an ab-sequence, balanced, two-phase set, where f_{as} and f_{bs} may be voltages, currents, or flux linkages, and f_s may be a function of time. For balanced steady-state conditions

$$F_{as} = \sqrt{2} F_s \cos[\omega_e t + \theta_{esf}(0)] \tag{3.5-3}$$

$$F_{bs} = \sqrt{2} F_s \sin[\omega_e t + \theta_{esf}(0)] \tag{3.5-4}$$

where F_s, ω_e, and $\theta_{esf}(0)$ are all constants. The subscript f is used to denote the zero displacement of the voltage $\theta_{esv}(0)$, current $\theta_{esi}(0)$, and flux linkage $\theta_{es\lambda}(0)$. In phasor form

$$\tilde{F}_{as} = F_s / \underline{\theta_{esf}(0)} \tag{3.5-5}$$

$$\tilde{F}_{bs} = F_s / \underline{\theta_{esf}(0) - \frac{\pi}{2}}$$
$$= -j\tilde{F}_{as} \tag{3.5-6}$$

For balanced, steady-state conditions it is not necessary to consider both the as and bs phasors since the phasors of the two-phase set are orthogonal.

Transforming the balanced set given by (3.5-1) and (3.5-2) to the arbitrary reference frame using (3.2-9) yields

$$f_{qs} = \sqrt{2} f_s \cos(\theta_{esf} - \theta) \tag{3.5-7}$$

$$f_{ds} = -\sqrt{2} f_s \sin(\theta_{esf} - \theta) \tag{3.5-8}$$

Please note that (3.5-7) and (3.5-8) for the two-phase system are the same as the balanced three-phase system, (3.4-20) and (3.4-21). This is due to the $\frac{2}{3}$ factor in (3.4-8). Therefore, the material in this section also applies to a balanced three-phase system. Equations (3.5-7) and (3.5-8) are now (3.5-1) and (3.5-2), the sinusoidal physical (real-life) variables, viewed from the arbitrary reference frame.

Asynchronously Rotating Reference Frames

For balanced steady-state conditions $\theta_{esf} = \omega_e t + \theta_{esf}(0)$, and since $\theta = \omega t + \theta(0)$, in the steady state, (3.5-7) and (3.5-8) become

$$F_{qs} = \sqrt{2} F_s \cos[(\omega_e - \omega)t + \theta_{esf}(0) - \theta(0)] \tag{3.5-9}$$

$$F_{ds} = -\sqrt{2} F_s \sin[(\omega_e - \omega)t + \theta_{esf}(0) - \theta(0)] \tag{3.5-10}$$

We now understand that the change of variables only changes the frequency of the balanced set. It does not change the amplitude of the balanced two-phase set, it only changes the speed and direction at which the magnetic field is rotating relative to us. This makes sense, since the transformation is nothing more than changing the speed we are rotating when we view Tesla's rotating magnetic field.

An asynchronously rotating reference frame occurs when $\omega \neq \omega_e$. When $\omega < \omega_e$, $(\omega_e - \omega)$ is positive and the mmf$_s$ is moving counterclockwise for $\tilde{I}_{as} = j\tilde{I}_{bs}$. Now, when $\omega > \omega_e$, $(\omega_e - \omega)$ becomes negative and the mmf$_s$ is moving clockwise for $\tilde{I}_{as} = j\tilde{I}_{bs}$; however, with $(\omega_e - \omega)$ negative, $e^{j(\omega_e - \omega)t}$ is rotating clockwise, and phasors are being rotated clockwise rather than counterclockwise which occurs when $(\omega_e - \omega)$ is positive, i.e. when $\omega < \omega_e$. In this case, because the sign of $\theta_{esi}(0)$ does not change, the current that appears lagging when viewed from a reference frame where $\omega < \omega_e$ becomes leading when viewed from a reference frame where $\omega > \omega_e$.

The direction of rotation of mmf, as viewed from a reference frame rotating at an angular velocity counterclockwise greater than ω_e, appears to be rotating clockwise at $|\omega_e - \omega|$ relative to us, toward synchronous speed. For $\omega < \omega_e$, mmf_s appears to be rotating counterclockwise at $|\omega_e - \omega|$, toward synchronous speed. We wonder what causes this to happen. Well, the reference frame variables (substitute variables) associated with the fictitious windings will provide us the correct view of mmfs that we would see from a given frame of reference. Let us see if this is the case. When $\omega < \omega_e$, F_{qs} (3.5-9) is a positive cosine and F_{ds} (3.5-10) is a negative sine wave. In this case, mmf_s appears to be rotating counterclockwise. Assume that I_{qs} is a positive cosine and I_{ds} is a negative sine in Figure 3.5-1. We see that the mmf_s is rotating counterclockwise. Now, when $\omega > \omega_e$, I_{qs} is still a cosine but now I_{ds} is a positive sine wave. Using Figure 3.5-1 again, we see that mmf_s is moving clockwise relative to us when we are moving faster counterclockwise than ω_e. Equation (2.2-12) tells us that the mmf_s is rotating clockwise; however, now we see that the qs and ds variables provide the appropriate view of Tesla's rotating magnetic field for $\omega < \omega_e$ and $\omega > \omega_e$.

Synchronously Rotating Reference Frame

In the synchronous reference frame, $\omega = \omega_e$, and with $\theta(0) = 0$, $\theta = \omega_e t$. For steady-state operation, (3.5-9) and (3.5-10) become

$$F_{qs}^e = \sqrt{2}\, F_s \cos \theta_{esf}(0) \tag{3.5-11}$$

$$F_{ds}^e = -\sqrt{2}\, F_s \sin \theta_{esf}(0) \tag{3.5-12}$$

The raised index e signifies variables in the synchronously reference frame. The as-phasor given by (3.5-5) can be written as

$$\tilde{F}_{as} = F_s \cos \theta_{esf}(0) + jF_s \sin \theta_{esf}(0) \tag{3.5-13}$$

which can be expressed in terms of F_{qs}^e and F_{ds}^e as

$$\sqrt{2}\, \tilde{F}_{as} = F_{qs}^e - j F_{ds}^e \tag{3.5-14}$$

We must be careful here; F_{qs}^e and F_{ds}^e are constants in the steady state. Clearly, $\tilde{F}_{as}$ is a complex number, and if $\omega = \omega_e$ and $\theta(0) = 0$, then $\tilde{F}_{as}$ can be expressed by (3.5-14).

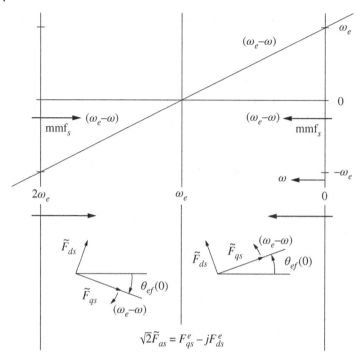

Figure 3.5-1 Direction of rotation of mmf$_s$ and the phase relation between $\tilde{F}_{qs}$ and $\tilde{F}_{ds}$ as viewed from the arbitrary reference frame when $\omega < \omega_e$ and $\omega > \omega_e$ for balanced steady-state conditions. For winding axes shown in Figure 3.2-1, $\tilde{I}_{as} = j\tilde{I}_{bs}$.

This is summarized in Figure 3.5-1, where we have shown the phasors of F_{qs} and F_{ds} and their relationship for $\omega < \omega_e$ and $\omega > \omega_e$. The rotating magnetic field viewed from any reference frame appears to be rotating toward synchronous speed.

For steady-state balanced conditions, the qs and ds variables in the synchronously rotating reference frame are constants; therefore, since pi^e_{qs} and pi^e_{ds} are zero and with $\omega = \omega_e$, we can write

$$V^e_{qs} = r_s I^e_{qs} + \omega_e L_{ss} I^e_{ds} \tag{3.5-15}$$

$$V^e_{ds} = r_s I^e_{ds} - \omega_e L_{ss} I^e_{qs} \tag{3.5-16}$$

From (3.4-14)

$$\tilde{V}_{as} = \frac{1}{\sqrt{2}}(V^e_{qs} - jV^e_{ds})$$

$$= (r_s + j\omega_e L_{ss})\tilde{I}_{as} \tag{3.5-17}$$

3.6 Closing Comments

We have established the link between Tesla's rotating magnetic field and reference frame theory. This provides a basis for Park's transformation as well as other change of variables set forth by various authors over the past 90 years. We have also established the voltage equations in the arbitrary reference frame for symmetrical stationary and rotating circuits. The material in this chapter solves a 90-year mystery started by Park in 1929. Reference frame theory is now much more understandable, and using it in machine analysis has become less mysterious.

References

1 Krause, P.C., Wasynczuk, O., O'Connell, T.C., and Hasan, M. (2018). Tesla's contribution to electric machine analysis. IEEE Paper, 2018 Summer Meeting (5–9 August 2018), Portland, OR

2 Krause, P.C., Wasynczuk, O., O'Connell, T.C., and Hasan, M. (2017). *Introduction to Power and Drive Systems*. New York: Wiley, IEEE Press.

3 Park, R.H. (1929). Two-reaction theory of synchronous machines – generalized method of analysis – part I. *AIEE Trans.* 48: 716–727.

4

Equivalent Circuits for the Symmetrical Machine

4.1 Introduction

The symmetrical machine has a symmetrical stator and rotor. It is commonly known as the induction machine which is the workhorse of the electrical system. We have introduced this type of machine in previous chapters but stopped short of transforming the flux-linkage equations to the arbitrary reference frame. This chapter is devoted to starting at this point and ending with the single-phase equivalent circuit from which steady-state operation of an induction machine may be determined.

4.2 Flux-Linkage Equations for a Magnetically Linear Two-Phase Symmetrical Machine

A two-phase symmetrical machine is shown in Figure 4.2-1. The flux-linkage equations for a magnetically linear machine may be expressed as

$$\begin{bmatrix} \lambda_{abs} \\ \lambda_{abr} \end{bmatrix} = \begin{bmatrix} \mathbf{L}_s & \mathbf{L}_{sr} \\ (\mathbf{L}_{sr})^T & \mathbf{L}_r \end{bmatrix} \begin{bmatrix} \mathbf{i}_{abs} \\ \mathbf{i}_{abr} \end{bmatrix} \tag{4.2-1}$$

where

$$\mathbf{L}_s = \begin{bmatrix} L_{ss} & 0 \\ 0 & L_{ss} \end{bmatrix} \tag{4.2-2}$$

$$\mathbf{L}_r = \begin{bmatrix} L_{rr} & 0 \\ 0 & L_{rr} \end{bmatrix} \tag{4.2-3}$$

$$L_{ss} = L_{ls} + L_{ms} \tag{4.2-4}$$

$$L_{rr} = L_{lr} + L_{mr} \tag{4.2-5}$$

Reference Frame Theory: Development and Applications, First Edition. Paul C. Krause.
© 2021 John Wiley & Sons, Inc. Published 2021 by John Wiley & Sons, Inc.

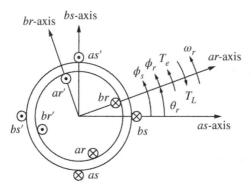

Figure 4.2-1 A two-pole two-phase symmetrical machine.

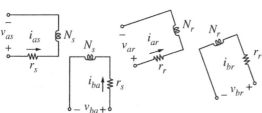

The inductances L_{ss} and L_{rr} each contain a leakage and a magnetizing inductance. The stator to rotor mutual inductances are expressed as

$$\mathbf{L}_{sr} = L_{sr} \begin{bmatrix} \cos\theta_r & -\sin\theta_r \\ \sin\theta_r & \cos\theta_r \end{bmatrix} \tag{4.2-6}$$

The rotor variables are referred to the stator windings by

$$\mathbf{i}'_{abr} = \frac{N_r}{N_s}\mathbf{i}_{abr} \tag{4.2-7}$$

$$\mathbf{v}'_{abr} = \frac{N_s}{N_r}\mathbf{v}_{abr} \tag{4.2-8}$$

$$\boldsymbol{\lambda}'_{abr} = \frac{N_s}{N_r}\boldsymbol{\lambda}_{abr} \tag{4.2-9}$$

where upon

$$\mathbf{v}_{abs} = \mathbf{r}_s\mathbf{i}_{abs} + p\boldsymbol{\lambda}_{abs} \tag{4.2-10}$$

$$\mathbf{v}'_{abr} = \mathbf{r}'_r\mathbf{i}'_{abr} + p\boldsymbol{\lambda}'_{abr} \tag{4.2-11}$$

and

$$\mathbf{r}'_r = \left(\frac{N_s}{N_r}\right)^2 \mathbf{r}_r \tag{4.2-12}$$

Now

$$\begin{bmatrix} \lambda_{abs} \\ \lambda'_{abr} \end{bmatrix} = \begin{bmatrix} \mathbf{L}_s & \mathbf{L}'_{sr} \\ (\mathbf{L}'_{sr})^T & \mathbf{L}'_r \end{bmatrix} \begin{bmatrix} \mathbf{i}_{abs} \\ \mathbf{i}'_{abr} \end{bmatrix}$$

(4.2-13)

where

$$\mathbf{L}'_r = \left(\frac{N_s}{N_r}\right)^2 \mathbf{L}_r = \begin{bmatrix} L'_{rr} & 0 \\ 0 & L'_{rr} \end{bmatrix}$$

(4.2-14)

where

$$L'_{lr} = \left(\frac{N_s}{N_r}\right)^2 L_{lr}$$

(4.2-15)

$$L_{ms} = \left(\frac{N_s}{N_r}\right)^2 L_{mr}$$

(4.2-16)

Therefore,

$$L'_{rr} = L'_{lr} + L_{ms}$$

(4.2-17)

and

$$\mathbf{L}'_{sr} = \frac{N_s}{N_r} \mathbf{L}_{sr}$$

$$= L_{ms} \begin{bmatrix} \cos\theta_r & -\sin\theta_r \\ \sin\theta_r & \cos\theta_r \end{bmatrix}$$

(4.2-18)

4.3 Flux-Linkage Equations in the Arbitrary Reference Frame

The transformation of the flux-linkage equations is

$$\lambda_{qds} = \mathbf{K}_s \lambda_{abs}$$

(4.3-1)

$$\lambda'_{qdr} = \mathbf{K}_r \lambda'_{abr}$$

(4.3-2)

which can be written as

$$\begin{bmatrix} \lambda_{qds} \\ \lambda'_{qdr} \end{bmatrix} = \begin{bmatrix} \mathbf{K}_s \mathbf{L}_s (\mathbf{K}_s)^{-1} & \mathbf{K}_s \mathbf{L}'_{sr} (\mathbf{K}_r)^{-1} \\ \mathbf{K}_r (\mathbf{L}'_{sr})^T (\mathbf{K}_s)^{-1} & \mathbf{K}_r \mathbf{L}'_r (\mathbf{K}_r)^{-1} \end{bmatrix} \begin{bmatrix} \mathbf{i}_{qds} \\ \mathbf{i}'_{qdr} \end{bmatrix}$$

(4.3-3)

Since $\mathbf{L}_s$ is given by (4.2-2) and $\mathbf{L}'_r$ by (4.2-14)

$$\mathbf{K}_s \mathbf{L}_s (\mathbf{K}_s)^{-1} = \mathbf{L}_s$$

(4.3-4)

$$\mathbf{K}_r \mathbf{L}'_r (\mathbf{K}_r)^{-1} = \mathbf{L}'_r$$

(4.3-5)

Also,

$$\mathbf{K}_s \mathbf{L}'_{sr} (\mathbf{K}_r)^{-1} = \mathbf{K}_r (\mathbf{L}'_{sr})^T (\mathbf{K}_s)^{-1} = \mathbf{L}_{ms} \tag{4.3-6}$$

where

$$\mathbf{L}_{ms} = \begin{bmatrix} L_{ms} & 0 \\ 0 & L_{ms} \end{bmatrix} \tag{4.3-7}$$

In expanded form

$$\lambda_{qs} = L_{ls} i_{qs} + L_{ms}(i_{qs} + i'_{qr}) \tag{4.3-8}$$

$$\lambda_{ds} = L_{ls} i_{ds} + L_{ms}(i_{ds} + i'_{dr}) \tag{4.3-9}$$

$$\lambda'_{qr} = L'_{lr} i'_{qr} + L_{ms}(i_{qs} + i'_{qr}) \tag{4.3-10}$$

$$\lambda'_{dr} = L'_{lr} i'_{dr} + L_{ms}(i_{ds} + i'_{dr}) \tag{4.3-11}$$

The equivalent circuit in the arbitrary reference frame is shown in Figure 4.3-1. The voltage equations in the arbitrary reference frame are

$$v_{qs} = r_s i_{qs} + \omega \lambda_{ds} + p\lambda_{qs} \tag{4.3-12}$$

$$v_{ds} = r_s i_{ds} - \omega \lambda_{qs} + p\lambda_{ds} \tag{4.3-13}$$

$$v'_{qr} = r'_r i'_{qr} + (\omega - \omega_r)\lambda'_{dr} + p\lambda'_{qr} \tag{4.3-14}$$

$$v'_{dr} = r'_r i'_{dr} - (\omega - \omega_r)\lambda'_{qr} + p\lambda'_{dr} \tag{4.3-15}$$

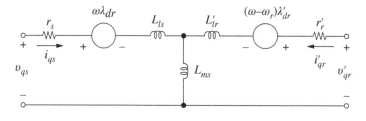

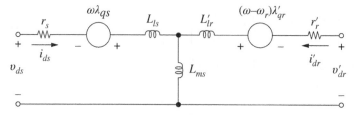

Figure 4.3-1 Arbitrary reference frame equivalent circuits for a two-phase, symmetrical machine.

4.4 Torque Expression in Arbitrary Reference Frame

The torque may be evaluated from

$$T_e(\mathbf{i}, \theta_r) = \frac{P}{2}\frac{\partial W_c(\mathbf{i}, \theta_r)}{\partial \theta_r} \tag{4.4-1}$$

In a magnetically linear system, the energy in the coupling field W_f and the coenergy W_c are equal. The field energy can be expressed as

$$W_f(\mathbf{i}, \theta_r) = \frac{1}{2}L_{ss}i_{as}^2 + \frac{1}{2}L_{ss}i_{bs}^2 + \frac{1}{2}L'_{rr}i'^2_{ar} + \frac{1}{2}L'_{rr}i'^2_{br} + L_{ms}i_{as}i'_{ar}\cos\theta_r$$
$$- L_{ms}i_{as}i'_{br}\sin\theta_r + L_{ms}i_{bs}i'_{ar}\sin\theta_r + L_{ms}i_{bs}i'_{br}\cos\theta_r \tag{4.4-2}$$

Since $W_f = W_c$, substituting (4.4-2) into (4.4-1) yields the electromagnetic torque for a magnetically linear P-pole two-phase symmetrical induction machine. In particular,

$$T_e = -\frac{P}{2}L_{ms}[(i_{as}i'_{ar} + i_{bs}i'_{br})\sin\theta_r + (i_{as}i'_{br} - i_{bs}i'_{ar})\cos\theta_r] \tag{4.4-3}$$

The torque and rotor speed are related by

$$T_e = J\frac{d\omega_{rm}}{dt} + B_m\omega_{rm} + T_L \tag{4.4-4}$$

where ω_{rm} is the actual rotor speed. For a P-pole machine, since $\omega_{rm} = (2/P)\omega_r$,

$$T_e = J\frac{2}{P}\frac{d\omega_r}{dt} + B_m\frac{2}{P}\omega_r + T_L \tag{4.4-5}$$

where J is the inertia of the rotor and, in some cases, the shaft connected load. The first term on the right-hand side is the inertial torque. The units of J are kilogram meter^2 (kg m^2) or joules second^2 (J s^2).

Equation (4.4-3) is an expression for the electromagnetic torque in machine currents i_{as}, i_{bs}, i'_{ar}, and i'_{br} for a magnetically linear system. If we express the torque in terms of the arbitrary reference frame currents, the electromagnetic torque for a magnetically linear system may be expressed as

$$T_e = \frac{P}{2}L_{ms}(i_{qs}i'_{dr} - i_{ds}i'_{qr}) \tag{4.4-6}$$

where T_e is positive for motor action. The equivalent expressions for torque are

$$T_e = \frac{P}{2}(\lambda'_{qr}i'_{dr} - \lambda'_{dr}i'_{qr}) \tag{4.4-7}$$

$$T_e = \frac{P}{2}(\lambda_{ds}i_{qs} - \lambda_{qs}i_{ds}) \tag{4.4-8}$$

When T_e is expressed in terms of flux linkages as in (4.4-7) and (4.4-8), it appears as if the leakage inductances play a role in the evaluation of torque. However, the product of leakage inductances times the currents cancels when the flux linkages are expressed in terms of inductances. It should also be mentioned that (4.4-7) and (4.4-8) are valid for magnetically linear and nonlinear systems.

4.5 Instantaneous and Steady-State Phasors

In Park's original paper [1], he likened the orthogonal q and d axes to a rotating complex plane and expressed the instantaneous q and d variables in complex form, in the synchronous reference frame. This expression is

$$\tilde{f}_{as}(t) = f_{qs}^e(t) - jf_{ds}^e(t) \qquad (4.5\text{-}1)$$

where $\tilde{f}_{as}(t)$ is the *instantaneous phasor* of the as-phase voltage, current, or flux linkage, and since it is in the synchronous reference frame it is rotating at ω_e counterclockwise as a phasor.

We know that a steady-state sinusoidal variable can be expressed as

$$F_a = F_p \cos \theta_{esf} \qquad (4.5\text{-}2)$$

which can be written as

$$F_a = \text{Re}[\sqrt{2}\tilde{F}_a e^{j\omega_e t}] \qquad (4.5\text{-}3)$$

where $\tilde{F}_a$ is a constant complex expression that is related to steady-state $\overset{e}{qs}$ and $\overset{e}{ds}$ quantities by

$$\sqrt{2}\tilde{F}_{as} = F_{qs}^e - jF_{ds}^e \qquad (4.5\text{-}4)$$

Since $f_{qs}^e(t)$ and $f_{ds}^e(t)$ are rotating at ω_e, we can write $f_{as}(t)$ in a form similar to (4.5-4) as

$$
\begin{aligned}
f_{as}(t) &= \text{Re}[\tilde{f}_{as}(t)e^{j\omega_e t}] \\
&= \text{Re}\{[f_{qs}^e(t) - jf_{ds}^e(t)]e^{j\omega_e t}\}
\end{aligned}
\qquad (4.5\text{-}5)
$$

We have referred to $\tilde{f}_{as}(t)$ as an instantaneous phasor, and since we are dealing with instantaneous quantities the $\sqrt{2}$ is not included as it was in (4.5-4) when using the rms values of steady-state sinusoidal variables.

The instantaneous phasor equations for the symmetrical machine may be obtained by substituting the v_{qs}^e and v_{ds}^e voltage equations into

$$\tilde{f}_{as} = f_{qs}^e - jf_{ds}^e \qquad (4.5\text{-}6)$$

$$\tilde{f}_{ar}' = f_{qr}'^e - jf_{dr}'^e \qquad (4.5\text{-}7)$$

where (4.5-6) is a repetition of (4.5-1). The voltage equations in the synchronous reference frame are from (4.3-12) through (4.3-15) with $\omega = \omega_e$.

$$v_{qs}^e = r_s i_{qs}^e + \omega_e \lambda_{ds}^e + p\lambda_{qs}^e \qquad (4.5\text{-}8)$$

$$v_{ds}^e = r_s i_{ds}^e - \omega_e \lambda_{qs}^e + p\lambda_{ds}^e \qquad (4.5\text{-}9)$$

$$v'^e_{qr} = r'_r i'^e_{qr} + (\omega_e - \omega_r)\lambda'^e_{dr} + p\lambda'^e_{qr} \tag{4.5-10}$$

$$v'^e_{dr} = r'_r i'^e_{dr} - (\omega_e - \omega_r)\lambda'^e_{qr} + p\lambda'^e_{dr} \tag{4.5-11}$$

Substituting (4.3-8) through (4.3-11) into (4.5-8) through (4.5-11) and then substituting (4.5-8) through (4.5-11) into (4.5-6) and (4.5-7) yields

$$\tilde{v}_{as} = r_s \tilde{i}_{as} + j\omega_e L_{ss}\tilde{i}_{as} + j\omega_e L_{ms}\tilde{i}'_{ar} + pL_{ss}\tilde{i}_{as} + pL_{ms}\tilde{i}'_{ar} \tag{4.5-12}$$

$$\tilde{v}'_{ar} = r'_r \tilde{i}'_{ar} + j(\omega_e - \omega_r)L'_{rr}\tilde{i}'_{ar} + j(\omega_e - \omega_r)L_{ms}\tilde{i}_{as} + pL'_{rr}\tilde{i}'_{ar} + pL_{ms}\tilde{i}_{as} \tag{4.5-13}$$

For steady-state conditions, the last two terms of (4.5-12) and (4.5-13) become zero, and we obtain the following steady-state voltage equations:

$$\tilde{V}_{as} = r_s \tilde{I}_{as} + j\omega_e(L_{ls} + L_{ms})\tilde{I}_{as} + j\omega_e L_{ms}\tilde{I}'_{ar} \tag{4.5-14}$$

$$\tilde{V}'_{ar} = r'_r \tilde{I}'_{ar} + j(\omega_e - \omega_r)(L'_{lr} + L_{ms})\tilde{I}'_{ar} + j(\omega_e - \omega_r)L_{ms}\tilde{I}_{as} \tag{4.5-15}$$

The so-called *slip* is

$$s = \frac{\omega_e - \omega_r}{\omega_e} \tag{4.5-16}$$

If we divide (4.5-15) by the slip, it becomes

$$\frac{\tilde{V}'_{ar}}{s} = \frac{r'_r}{s}\tilde{I}'_{ar} + j\omega_e(L'_{lr} + L_{ms})\tilde{I}'_{ar} + j\omega_e L_{ms}\tilde{I}_{as} \tag{4.5-17}$$

Equations (4.5-14) and (4.5-17) suggest the single-phase equivalent T circuit of a two-phase symmetrical induction machine during steady-state balanced operation shown in Figure 4.5-1.

An expression for the steady-state electromagnetic torque may be obtained by first writing (4.4-6) in terms of I^e_{qs}, I^e_{ds}, I'^e_{qr}, and I'^e_{dr} and then using (4.5-6) and (4.5-7) to express $\tilde{I}_{as}$ and $\tilde{I}'_{ar}$. The expression may be reduced to

$$T_e = 2\left(\frac{P}{2}\right)L_{ms}\text{Re}[j\tilde{I}^*_{as}\tilde{I}'_{ar}] \tag{4.5-18}$$

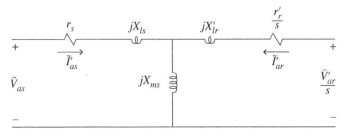

Figure 4.5-1 Equivalent circuit for a two-phase symmetrical induction machine for balanced steady-state operation.

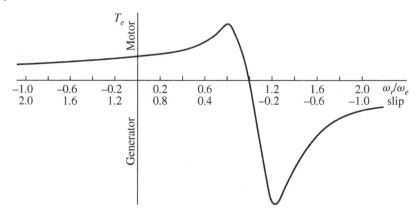

Figure 4.5-2 Steady-state torque–speed characteristics of a symmetrical induction machine.

where $\tilde{I}_{as}^{*}$ is the conjugate of $\tilde{I}_{as}$. The phasor currents may be calculated from the equivalent circuit given in Figure 4.5-1.

The typical torque–speed characteristics of many single-fed two-phase induction or symmetrical machines are shown in Figure 4.5-2. The parameters of the machine are often selected so that, for rated frequency operation, the maximum torque occurs between 80 and 90% of synchronous speed. Generally, the maximum torque is two or three times the rated torque of the machine. Although we are considering a two-phase machine, the general shape of the torque–speed characteristic is similar for multiphase induction machines.

A single-fed two-pole two-phase 5-hp 110-V (rms) 60-Hz induction machine is shown in Figure 4.5-3, with the following parameters: $r_{s} = 0.295\,\Omega$, $L_{ls} = 0.944\,\text{mH}$, $L_{ms} = 35.15\,\text{mH}$, $r_{r}' = 0.201\,\Omega$, and $L_{lr}' = 0.944\,\text{mH}$. The phasor diagrams for $s = 0.05$, motor action, and $s = -0.05$, generator action, are shown in Figures 4.5-4 and 4.5-5, respectively. In Figure 4.5-4, the stator poles are "pushing" the rotor poles in the counterclockwise direction, motor action. In Figure 4.5-5, the rotor poles are "pushing" the stator poles counterclockwise, generator action.

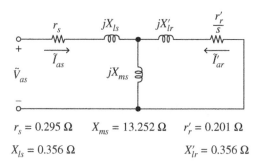

Figure 4.5-3 Equivalent circuit for steady-state operation of a single-fed symmetrical machine.

$r_{s} = 0.295\,\Omega$ $\quad X_{ms} = 13.252\,\Omega$ $\quad r_{r}' = 0.201\,\Omega$

$X_{ls} = 0.356\,\Omega$ $\qquad\qquad\qquad X_{lr}' = 0.356\,\Omega$

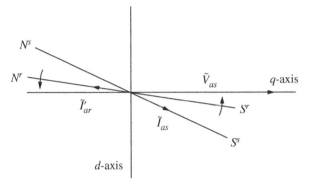

Figure 4.5-4   Phasor diagram; motor action, $s = 0.05$.

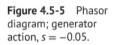

Figure 4.5-5   Phasor diagram; generator action, $s = -0.05$.

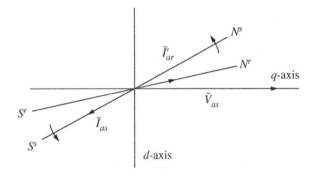

4.6   Flux-Linkage Equations for a Magnetically Linear Three-Phase Symmetrical Machine and Equivalent Circuit

A three-phase symmetrical machine is shown in Figure 4.6-1. Each stator phase has N_s equivalent turns; each rotor phase has N_r equivalent turns.

The flux linkage equations may be written as

$$\begin{bmatrix} \boldsymbol{\lambda}_{abcs} \\ \boldsymbol{\lambda}_{abcr} \end{bmatrix} = \begin{bmatrix} \mathbf{L}_s & \mathbf{L}_{sr} \\ (\mathbf{L}_{sr})^T & \mathbf{L}_r \end{bmatrix} \begin{bmatrix} \mathbf{i}_{abcs} \\ \mathbf{i}_{abcr} \end{bmatrix} \tag{4.6-1}$$

Although the same notation is used for the inductance matrices as for the two-phase case, they are not the same. The self-inductances are all constant and can be expressed as

$$\mathbf{L}_s = \begin{bmatrix} L_{ss} & -\dfrac{1}{2}L_{ms} & -\dfrac{1}{2}L_{ms} \\ -\dfrac{1}{2}L_{ms} & L_{ss} & -\dfrac{1}{2}L_{ms} \\ -\dfrac{1}{2}L_{ms} & -\dfrac{1}{2}L_{ms} & L_{ss} \end{bmatrix} \tag{4.6-2}$$

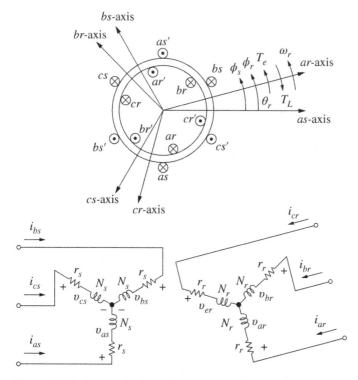

Figure 4.6-1 A two-pole three-phase symmetrical machine.

$$\mathbf{L}_r = \begin{bmatrix} L_{rr} & -\dfrac{1}{2}L_{mr} & -\dfrac{1}{2}L_{mr} \\[2mm] -\dfrac{1}{2}L_{mr} & L_{rr} & -\dfrac{1}{2}L_{mr} \\[2mm] -\dfrac{1}{2}L_{mr} & -\dfrac{1}{2}L_{mr} & L_{rr} \end{bmatrix} \tag{4.6-3}$$

where $L_{ss} = L_{ls} + L_{ms}$ and $L_{rr} = L_{lr} + L_{mr}$. Also,

$$\mathbf{L}_{sr} = L_{sr} \begin{bmatrix} \cos\theta_r & \cos\left(\theta_r + \dfrac{2}{3}\pi\right) & \cos\left(\theta_r - \dfrac{2}{3}\pi\right) \\[2mm] \cos\left(\theta_r - \dfrac{2}{3}\pi\right) & \cos\theta_r & \cos\left(\theta_r + \dfrac{2}{3}\pi\right) \\[2mm] \cos\left(\theta_r + \dfrac{2}{3}\pi\right) & \cos\left(\theta_r - \dfrac{2}{3}\pi\right) & \cos\theta_r \end{bmatrix} \tag{4.6-4}$$

where

$$L_{sr} = \frac{N_s N_r}{\mathscr{R}_m} \tag{4.6-5}$$

All rotor variables may be referred to the stator windings by the following turns ratios:

$$\mathbf{i}'_{abcr} = \frac{N_r}{N_s}\mathbf{i}_{abcr} \tag{4.6-6}$$

$$\mathbf{v}'_{abcr} = \frac{N_s}{N_r}\mathbf{v}_{abcr} \tag{4.6-7}$$

$$\boldsymbol{\lambda}'_{abcr} = \frac{N_s}{N_r}\boldsymbol{\lambda}_{abcr} \tag{4.6-8}$$

The flux linkage equations may now be written as

$$\begin{bmatrix} \boldsymbol{\lambda}_{abcs} \\ \boldsymbol{\lambda}'_{abcr} \end{bmatrix} = \begin{bmatrix} \mathbf{L}_s & \mathbf{L}'_{sr} \\ (\mathbf{L}'_{sr})^T & \mathbf{L}'_r \end{bmatrix} \begin{bmatrix} \mathbf{i}_{abcs} \\ \mathbf{i}'_{abcr} \end{bmatrix} \tag{4.6-9}$$

where, by definition,

$$\mathbf{L}'_{sr} = \frac{N_s}{N_r}\mathbf{L}_{sr} = \frac{L_{ms}}{L_{sr}}\mathbf{L}_{sr} \tag{4.6-10}$$

and

$$\mathbf{L}'_r = \begin{bmatrix} L'_{lr} + L_{ms} & -\frac{1}{2}L_{ms} & -\frac{1}{2}L_{ms} \\ -\frac{1}{2}L_{ms} & L'_{lr} + L_{ms} & -\frac{1}{2}L_{ms} \\ -\frac{1}{2}L_{ms} & -\frac{1}{2}L_{ms} & L'_{lr} + L_{ms} \end{bmatrix} \tag{4.6-11}$$

In (4.6-11),

$$L'_{lr} = \left(\frac{N_s}{N_r}\right)^2 L_{lr} \tag{4.6-12}$$

A change of variables that formulates a transformation of the three-phase variables of stationary (stator) circuits to the arbitrary reference frame may by expressed as

$$\mathbf{f}_{qd0s} = \mathbf{K}_s\mathbf{f}_{abcs} \tag{4.6-13}$$

where

$$(\mathbf{f}_{qd0s})^T = [f_{qs} \ f_{ds} \ f_{0s}] \tag{4.6-14}$$

$$(\mathbf{f}_{abcs})^T = [f_{as} \ f_{bs} \ f_{cs}] \tag{4.6-15}$$

$$\mathbf{K}_s = \frac{2}{3}\begin{bmatrix} \cos\theta & \cos\left(\theta - \frac{2}{3}\pi\right) & \cos\left(\theta + \frac{2}{3}\pi\right) \\ \sin\theta & \sin\left(\theta - \frac{2}{3}\pi\right) & \sin\left(\theta + \frac{2}{3}\pi\right) \\ \frac{1}{2} & \frac{1}{2} & \frac{1}{2} \end{bmatrix} \tag{4.6-16}$$

where the $\frac{2}{3}$ factor was introduced by Park [1].

$$\frac{d\theta}{dt} = \omega(t) \tag{4.6-17}$$

It can be shown that the inverse is

$$(\mathbf{K}_s)^{-1} = \begin{bmatrix} \cos\theta & \sin\theta & 1 \\ \cos\left(\theta - \frac{2}{3}\pi\right) & \sin\left(\theta - \frac{2}{3}\pi\right) & 1 \\ \cos\left(\theta + \frac{2}{3}\pi\right) & \sin\left(\theta + \frac{2}{3}\pi\right) & 1 \end{bmatrix} \tag{4.6-18}$$

The s subscript indicates the variables, parameters, and transformation associated with stationary circuits.

A change of variables which formulates a transformation of the three-phase variables of the rotor circuits to the arbitrary reference frame is

$$\mathbf{f}'_{qd0r} = \mathbf{K}_r \mathbf{f}'_{abcr} \tag{4.6-19}$$

where

$$(\mathbf{f}'_{qd0r})^T = [f'_{qr} \ f'_{dr} \ f'_{0r}] \tag{4.6-20}$$

$$(\mathbf{f}'_{abcr})^T = [f'_{ar} \ f'_{br} \ f'_{cr}] \tag{4.6-21}$$

$$\mathbf{K}_r = \frac{2}{3} \begin{bmatrix} \cos\beta & \cos\left(\beta - \frac{2}{3}\pi\right) & \cos\left(\beta + \frac{2}{3}\pi\right) \\ \sin\beta & \sin\left(\beta - \frac{2}{3}\pi\right) & \sin\left(\beta + \frac{2}{3}\pi\right) \\ \frac{1}{2} & \frac{1}{2} & \frac{1}{2} \end{bmatrix} \tag{4.6-22}$$

$$\beta = \theta - \theta_r \tag{4.6-23}$$

where the angular position is defined by

$$\frac{d\beta}{dt} = \omega(t) - \omega_r(t) \tag{4.6-24}$$

The inverse is

$$(\mathbf{K}_r)^{-1} = \begin{bmatrix} \cos\beta & \sin\beta & 1 \\ \cos\left(\beta - \frac{2}{3}\pi\right) & \sin\left(\beta - \frac{2}{3}\pi\right) & 1 \\ \cos\left(\beta + \frac{2}{3}\pi\right) & \sin\left(\beta + \frac{2}{3}\pi\right) & 1 \end{bmatrix} \tag{4.6-25}$$

The r subscript indicates the variables, parameters, and transformation associated with rotating circuits. Transforming the voltage equations yields

$$v_{qs} = r_s i_{qs} + \omega\lambda_{ds} + p\lambda_{qs} \tag{4.6-26}$$

$$v_{ds} = r_s i_{ds} - \omega \lambda_{qs} + p\lambda_{ds} \tag{4.6-27}$$

$$v_{0s} = r_s i_{0s} + p\lambda_{0s} \tag{4.6-28}$$

$$v'_{qr} = r'_r i'_{qr} + (\omega - \omega_r)\lambda'_{dr} + p\lambda'_{qr} \tag{4.6-29}$$

$$v'_{dr} = r'_r i'_{dr} - (\omega - \omega_r)\lambda'_{qr} + p\lambda'_{dr} \tag{4.6-30}$$

$$v'_{0r} = r'_r i'_{0r} + p\lambda'_{0r} \tag{4.6-31}$$

Transforming (4.6-9) yields

$$\lambda_{qs} = L_{ls} i_{qs} + L_{Ms}(i_{qs} + i'_{qr}) \tag{4.6-32}$$

$$\lambda_{ds} = L_{ls} i_{ds} + L_{Ms}(i_{ds} + i'_{dr}) \tag{4.6-33}$$

$$\lambda_{0s} = L_{ls} i_{0s} \tag{4.6-34}$$

$$\lambda'_{qr} = L'_{lr} i'_{qr} + L_{Ms}(i_{qs} + i'_{qr}) \tag{4.6-35}$$

$$\lambda'_{dr} = L'_{lr} i'_{dr} + L_{Ms}(i_{ds} + i'_{dr}) \tag{4.6-36}$$

$$\lambda'_{0r} = L'_{lr} i'_{0r} \tag{4.6-37}$$

where $L_{Ms} = \frac{3}{2} L_{ms}$.

Equations (4.6-26) through (4.6-37) suggest the equivalent circuits shown in Figure 4.6-2 [2].

Other than the zero-voltage equations and the $\frac{3}{2}$ factor, the voltage equations are the same as those for a two-phase machine. Moreover, if the three-phase machine is connected in wye without a neutral connection, as shown in Figure 4.6-1, the currents i_{0s} and i'_{0r} are zero for balanced or unbalanced operation since the sum of the three-phase currents is zero. Therefore, v_{0s} and v'_{0r} are zero since the sum of the three-phase stator and rotor flux linkages will be zero for symmetrical systems. It is clear that the work and single-phase equivalent given in Section 4.5 is valid for a three-phase machine if we replace L_{ms} with L_{Ms}. Also, the torque expression must be multiplied by $\frac{3}{2}$.

4.7 Closing Comments

We have set forth the equivalent circuits for the two- and three-phase symmetrical machine. We have found that the equivalent circuits involve a $\frac{3}{2}$ times the magnetizing inductance L_{ms}, and the expressions for torque must be multiplied by $\frac{3}{2}$ for

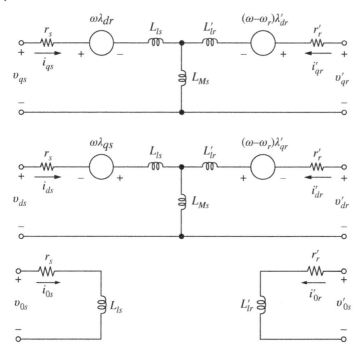

Figure 4.6-2 Arbitrary reference frame equivalent circuits for a three-phase symmetrical machine.

the three-phase machine. We also modify the equivalent circuits to establish the equivalent circuits for other machines that we will consider in later chapters.

References

1 Park, R.H. (1929). Two-reaction theory of synchronous machines – generalized method of analysis – part I. *AIEE Trans.* 48: 716–727.
2 Krause, P.C. and Thomas, C.H. (1965). Simulation of symmetrical induction machinery. *IEEE Trans. Power App. Syst.* 84: 1038–1053.

5

Synchronous Machines

5.1 Introduction

Synchronous machines have a symmetrical stator and an unsymmetrical rotor. They are referred to as synchronous machines since in the steady state the rotor must rotate in synchronous with the rotating magnetic field established by the stator. In the case of a power system this is generally fixed in magnitude and at 50 or 60 Hz.

5.2 Synchronous Machine

A three-phase synchronous generator is shown in Figure 5.2-1 [1]. The stator windings are identical, sinusoidally distributed windings. For analysis purposes, the electrical characteristics of the rotor of a synchronous machine may be adequately represented with a field winding (*fd* winding) and short-circuited *damper* or *amortisseur* windings (*kq* and *kd* windings). Although the damper windings are shown with provisions to apply a voltage, they are, in fact, short-circuited windings which represent the paths for induced rotor currents. We assume that the damper windings are approximated by two sinusoidally distributed windings displaced in space by 90° electrical degrees. The *kd* winding has the same magnetic axis as the *fd* winding; it has N_{kd} equivalent turns with resistance r_{kd}. The magnetic axis of *kq* winding is orthogonal with the magnetic axis of the *fd* and *kd* windings. It has N_{kq} equivalent turns and r_{kq} resistance. The rotor configuration shown in Figure 5.2-1 for a three-phase machine is the same for any multiphase two-pole synchronous machine. The quadrature axis (*q*-axis) and direct axis (*d*-axis) are also shown in Figure 5.2-1. The *q*-axis is the magnetic axis of the *kq* winding, whereas the *d*-axis is the magnetic axis of the *fd* and *kd* windings. In synchronous machine analysis, but not in general, the *q* and *d* axes are reserved to denote the rotor magnetic axes since, over the years, they

Reference Frame Theory: Development and Applications, First Edition. Paul C. Krause.

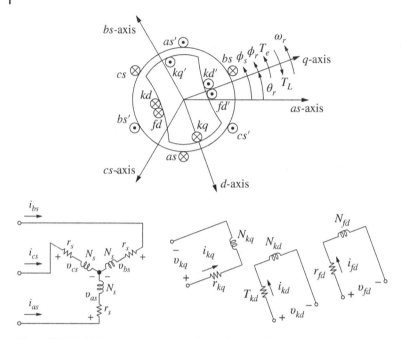

Figure 5.2-1 Salient-rotor two-pole three-phase salient-pole synchronous machine.

have been associated with the physical structure of the synchronous machine rotor quite independent of any transformation. However, as we have seen in the previous chapters, these need not be associated with any physical axes when using them for general machine analysis.

Albeit small, a reluctance torque is also developed at synchronous speed due to the nonuniform air gap as a result of the saliency of the rotor. The so-called salient-pole construction is common for slower speed machines (large number of poles) such as hydroturbine generators. In this type of rotor construction, the field winding is wound upon the rotor surface, as shown in Figure 5.2-1, and the air gap is nonuniform to make room for the placement of the field winding. Therefore, the q-axis magnetic path has a higher reluctance than the d-axis magnetic path.

It was found early on that a synchronous machine with only a field winding would tend to oscillate about synchronous speed in a slowly damped manner following any slight disturbance. Adding damper windings (short-circuited rotor windings) provided the desired damping by induction machine action. Currents are induced in these rotor (damper) windings whenever the speed of the rotor differs from the angular velocity of the rotating air-gap mmf established by the stator currents (mmf$_s$).

The main torque of a synchronous machine is developed at synchronous speed because of the interaction of mmf_s and mmf_r. At synchronous speed, current is not induced in the damper windings. The question arises as to why this oscillation does not occur in the brushless dc drive? The answer is the fact that the frequency and amplitude control of the applied stator voltages inherently prevents this oscillation from occurring.

Torque is torque by whatever means it is developed, and perhaps, we should not emphasize the separation of torque into three types (interaction of mmf_s and mmf_r, reluctance, and induction) since the dynamic operation of the machine is described by nonlinear equations, and superposition cannot be applied, in general. Nevertheless, this separation is helpful in understanding the operation of the synchronous machine.

5.3 Equivalent Circuit For Three-Phase Synchronous Generator

Rather than going through the long derivation of the equivalent circuit, we modify the work we have done for the symmetrical machine to fit the three-phase synchronous machine. The equivalent circuit for the three-phase synchronous machine in the rotor reference frame is shown in Figure 5.3-1. The damper windings are shown with an applied voltage; however, they are actually short-circuited squirrel cage-type bars. The reference frame is rotating at the angular velocity of the rotor; therefore, only the stator circuits are transformed.

The voltage equations become

$$v^r_{qs} = r_s i^r_{qs} + \omega_r \lambda^r_{ds} + p\lambda^r_{qs} \tag{5.3-1}$$

$$v^r_{ds} = r_s i^r_{ds} - \omega_r \lambda^r_{qs} + p\lambda^r_{ds} \tag{5.3-2}$$

$$v_{0s} = r_s i_{0s} + p\lambda_{0s} \tag{5.3-3}$$

$$v''^r_{kq} = r'_{kq} i''^r_{kq} + p\lambda''^r_{kq} \tag{5.3-4}$$

$$v''^r_{fd} = r'_{fd} i''^r_{fd} + p\lambda''^r_{fd} \tag{5.3-5}$$

$$v''^r_{kd} = r'_{kq} i''^r_{kd} + p\lambda''^r_{kd} \tag{5.3-6}$$

The flux-linkage equations become

$$\lambda^r_{qs} = L_{ls} i^r_{qs} + L_{Mq}(i^r_{qs} + i''^r_{kq}) \tag{5.3-7}$$

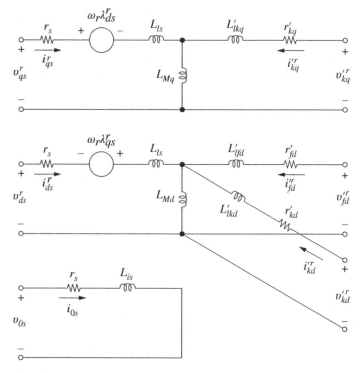

Figure 5.3-1 Equivalent circuit for three-phase synchronous machine in the rotor reference frame.

$$\lambda_{ds}^{r} = L_{ls}i_{ds}^{r} + L_{Md}(i_{ds}^{r} + i''^{r}_{fd} + i''^{r}_{kd}) \tag{5.3-8}$$

$$\lambda_{0s} = L_{ls}i_{0s} \tag{5.3-9}$$

$$\lambda''^{r}_{kq} = L'_{lkq}i''^{r}_{kq} + L_{Mq}(i_{qs}^{r} + i''^{r}_{kq}) \tag{5.3-10}$$

$$\lambda''^{r}_{fd} = L'_{lfd}i''^{r}_{fd} + L_{Md}(i_{ds}^{r} + i''^{r}_{fd} + i''^{r}_{kd}) \tag{5.3-11}$$

$$\lambda''^{r}_{kd} = L'_{lkd}i''^{r}_{kd} + L_{Md}(i_{ds}^{r} + i''^{r}_{fd} + i''^{r}_{kd}) \tag{5.3-12}$$

where $L_{Mq} = \frac{3}{2}L_{mq}$ and $L_{Md} = \frac{3}{2}L_{md}$.

In the above equations, the resistances and leakage inductances referred to the stator windings are

$$r_j = \frac{3}{2}\left(\frac{N_s}{N_j}\right)^2 r_j \tag{5.3-13}$$

$$L'_{\ell j} = \frac{3}{2}\left(\frac{N_s}{N_j}\right)^2 L_{\ell j} \tag{5.3-14}$$

where j may be kq, f_d, or kd.

During steady-state balanced conditions, the stator and the field windings are the only windings carrying current. The voltage equations become

$$V^r_{qs} = r_s I^r_{qs} + \omega_r \lambda^r_{ds} \tag{5.3-15}$$

$$V^r_{ds} = r_s I^r_{ds} - \omega_r \lambda^r_{qs} \tag{5.3-16}$$

$$V'^r_{fd} = r'_{fd} I'^r_{fd} \tag{5.3-17}$$

where

$$\lambda^r_{qs} = L_{ls} I^r_{qs} + L_{Mq} I^r_{qs} \tag{5.3-18}$$

$$\lambda^r_{ds} = L_{ls} I^r_{ds} + L_{Md}(I^r_{ds} + I'^r_{fd}) \tag{5.3-19}$$

Now, the rotor angle is defined as

$$
\begin{aligned}
\delta &= \theta_r - \theta_{esv} \\
&= \int_0^t [\omega_r(\xi) - \omega_e(\xi)]\, d\xi + \theta_r(0) - \theta_{esv}(0)
\end{aligned}
\tag{5.3-20}
$$

In the steady state, $\omega_r = \omega_e$ and δ is the angle between the q-axis and the phase angle of the *as* voltage, which is generally selected as zero. In general, however, δ in the steady state is

$$\delta = \theta_r(0) - \theta_{esv}(0) \tag{5.3-21}$$

Transforming a balanced steady-state three-phase set of *abc*-sequence yields

$$F^r_{qs} = \sqrt{2}F_s \cos[\theta_{esf}(0) - \theta_r(0)] \tag{5.3-22}$$

$$F^r_{ds} = -\sqrt{2}F_s \sin[\theta_{esf}(0) - \theta_r(0)] \tag{5.3-23}$$

Substituting $\theta_r(0)$ from (5.3-21) into (5.3-22) and (5.3-23) yields

$$F^r_{qs} = \sqrt{2}F_s \cos[\theta_{esf}(0) - \theta_{esv}(0) - \delta] \tag{5.3-24}$$

$$F^r_{ds} = -\sqrt{2}F_s \sin[\theta_{esf}(0) - \theta_{esv}(0) - \delta] \tag{5.3-25}$$

Now, $\tilde{F}_{as}$ may be written as

$$\tilde{F}_{as} = F_s e^{j\theta_{esf}(0)} \tag{5.3-26}$$

If we multiply (5.3-26) by $\sqrt{2}e^{-j\delta}$, and if we let $\theta_{esv}(0) = 0$ or in other words $\tilde{V}_{as} = V_s/\underline{0^\circ}$, we can write

$$\sqrt{2}\tilde{F}_{as}e^{-j\delta} = \sqrt{2}F_s \cos[\theta_{esf}(0) - \delta] + j\sqrt{2}F_s \sin[\theta_{esf}(0) - \delta] \tag{5.3-27}$$

Substituting (5.3-24) and (5.3-25) into (5.3-27) with $\theta_{esv}(0) = 0$, we can write

$$\sqrt{2}\tilde{F}_{as}e^{-j\delta} = F_{qs}^r - jF_{ds}^r \tag{5.3-28}$$

Substituting (5.3-15) and (5.3-16) into (5.3-28) yields

$$\sqrt{2}\tilde{V}_{as}e^{-j\delta} = r_s I_{qs}^r + X_d I_{ds}^r + \omega_e L_{Md} I_{fd}^{\prime r} + j(-r_s I_{ds}^r + X_q I_{qs}^r) \tag{5.3-29}$$

where $\frac{3}{2}X_{md} = \omega_e L_{Md}$.

Now

$$j\sqrt{2}\tilde{I}_{as}e^{-j\delta} = I_{ds}^r + jI_{qs}^r \tag{5.3-30}$$

In (5.3-29)

$$X_q = X_{ls} + X_{Mq} \tag{5.3-31}$$

$$X_d = X_{ls} + X_{Md} \tag{5.3-32}$$

If we add and substitute $X_q I_{ds}^r$ from the right-hand side of (5.3-29), we can write it as

$$\tilde{V}_{as} = (r_s + jX_q)\tilde{I}_{as} + \frac{1}{\sqrt{2}}[(X_d - X_q)I_{ds}^r + X_{Md}I_{fd}^{\prime r}]e^{j\delta} \tag{5.3-33}$$

The last term of (5.3-33) may be written as

$$\tilde{E}_a = \frac{1}{\sqrt{2}}[(X_d - X_q)I_{ds}^r + X_{Md}I_{fd}^{\prime r}]e^{j\delta} \tag{5.3-34}$$

We can now write (5.3-31) as

$$\tilde{V}_{as} = (r_s + jX_q)\tilde{I}_{as} + \tilde{E}_a \tag{5.3-35}$$

The phasor diagram is shown in Figure 5.3-2 for typical generator action. The rotor poles are "pushing" the stator poles counterclockwise.

The torque can be expressed as

$$T_e = \left(\frac{3}{2}\right)\left(\frac{P}{2}\right)(\lambda_{ds}^r i_{qs}^r - \lambda_{qs}^r i_{ds}^r) \tag{5.3-36}$$

The steady-state torque for balanced conditions may be written as

$$T_e = -\left(\frac{3}{2}\right)\left(\frac{P}{2}\right)\frac{1}{\omega_e}\left[\frac{\omega_e L_{Md} I_{fd}^{\prime r}\sqrt{2}V_s}{X_d}\sin\delta + \frac{1}{2}\left(\frac{1}{X_q} - \frac{1}{X_d}\right)(\sqrt{2}V_s)^2\sin 2\delta\right]$$

$$\tag{5.3-37}$$

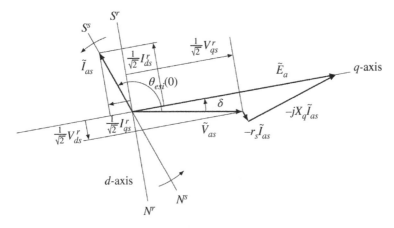

Figure 5.3-2 Phasor diagram for generator operation.

5.4 Closing Comment

The dynamic and steady-state voltage and torque equations have been set forth for the generator used in the electric grid.

Reference

1 Krause, P.C., Wasynczuk, O., and Sudhoff, S.D. (2002). *Analysis of Electric Machinery and Drive Systems*, 2e. IEEE Press, Wiley.

6

Brushless dc Drive with Field Orientation

6.1 Introduction

In the previous chapter we established the equations for a synchronous machine that would be used as a generator in a power grid where the voltage is controlled essentially constant and the frequency is also constant at 50 or 60 Hz. There is another synchronous machine used in a drive system: the brushless dc drive. It is a permanent-magnet ac machine that is driven by an inverter, generally a six-step inverter. The stator voltages are controlled by the inverter so that the frequency of the stator voltages is the same angular velocity as the rotor. This assures that the permanent magnet of the rotor is running in synchronism with mmf$_s$ established by the stator currents. As a torque load is applied to the shaft, the rotor slows and instantaneously the frequency of the applied voltages is decreased in unison with the rotor. This is done very accurately and very fast. This chapter is devoted to the brushless dc drive which gets its name due to the fact that this drive has torque speed characteristics of a permanent-magnet dc machine.

6.2 The Permanent-Magnet ac Machine

The two-phase permanent-magnet ac machine is shown in Figure 6.2-1. The stator is symmetrical with N_s turns in each phase winding. The voltage equations are

$$v_{as} = r_s i_{as} + \frac{d\lambda_{as}}{dt} \tag{6.2-1}$$

$$v_{bs} = r_s i_{bs} + \frac{d\lambda_{bs}}{dt} \tag{6.2-2}$$

The flux-linkage equations may be expressed as

$$\lambda_{as} = L_{asas} i_{as} + L_{asbs} i_{bs} + \lambda_{asm} \tag{6.2-3}$$

Reference Frame Theory: Development and Applications, First Edition. Paul C. Krause.
© 2021 John Wiley & Sons, Inc. Published 2021 by John Wiley & Sons, Inc.

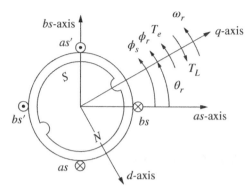

Figure 6.2-1 Two-pole two-phase permanent-magnet ac machine.

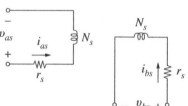

$$\lambda_{bs} = L_{bsas}i_{as} + L_{bsbs}i_{bs} + \lambda_{bsm} \tag{6.2-4}$$

which may be written as

$$\lambda_{abs} = \mathbf{L}_s\mathbf{i}_{abs} + \lambda'_m \tag{6.2-5}$$

where λ'_m is

$$\lambda'_m = \lambda'_m \begin{bmatrix} \sin\theta_r \\ -\cos\theta_r \end{bmatrix} \tag{6.2-6}$$

where λ'_m is the amplitude of the flux linkages established by the permanent magnet as viewed from the stator phase windings. It is proportional to the amplitude of the open circuit phase winding voltage. Also, θ_r in (6.2-6) is

$$\frac{d\theta_r}{dt} = \omega_r(t) \tag{6.2-7}$$

The inductance $\mathbf{L}_s$ is

$$\mathbf{L}_s = \begin{bmatrix} L_{ss} & 0 \\ 0 & L_{ss} \end{bmatrix} \tag{6.2-8}$$

where

$$L_{ss} = L_{ls} + L_{ms} \tag{6.2-9}$$

The voltage equations in the rotor reference frame are

$$v_{qs}^r = r_s i_{qs}^r + \omega_r \lambda_{ds}^r + p\lambda_{qs}^r \tag{6.2-10}$$

$$v_{ds}^r = r_s i_{ds}^r - \omega_r \lambda_{qs}^r + p\lambda_{ds}^r \tag{6.2-11}$$

The stator flux linkages given by (6.2-5) may be transformed to the rotor reference frame by

$$(\mathbf{K}_s^r)^{-1}\lambda_{qds}^r = \mathbf{L}_s(\mathbf{K}_s^r)^{-1}\mathbf{i}_{qds}^r + \lambda_m' \tag{6.2-12}$$

Substituting (6.2-6) for λ_m' and (6.2-8) for $\mathbf{L}_s$ we have

$$\lambda_{qs}^r = L_{ss} i_{qs}^r \tag{6.2-13}$$

$$\lambda_{ds}^r = L_{ss} i_{ds}^r + \lambda_m''^r \tag{6.2-14}$$

Since $p\lambda_m''^r = 0$, substituting (6.2-13) and (6.2-14) into (6.2-10) and (6.2-11) yields

$$v_{qs}^r = (r_s + pL_{ss})i_{qs}^r + \omega_r L_{ss} i_{ds}^r + \omega_r \lambda_m''^r \tag{6.2-15}$$

$$v_{ds}^r = (r_s + pL_{ss})i_{ds}^r - \omega_r L_{ss} i_{qs}^r \tag{6.2-16}$$

where $\omega_r = \omega_e$.

Equations (6.2-15) and (6.2-16) suggest the equivalent circuits shown in Figure 6.2-2. For balanced steady-state conditions, (6.2-15) and (6.2-16) become

$$V_{qs}^r = r_s I_{qs}^r + \omega_r L_{ss} I_{ds}^r + \omega_r \lambda_m''^r \tag{6.2-17}$$

$$V_{ds}^r = r_s I_{ds}^r - \omega_r L_{ss} I_{qs}^r \tag{6.2-18}$$

Since we are assuming a linear magnetic system, the coenergy may be expressed as

$$W_c = \frac{1}{2}L_{ss}(i_{as}^2 + i_{bs}^2) + \lambda_m' i_{as} \sin \theta_r - \lambda_m' i_{bs} \cos \theta_r + W_{pm} \tag{6.2-19}$$

where W_{pm} is the energy associated with the permanent magnet, which is assumed constant for the device shown in Figure 6.2-1. Taking the partial derivative of (6.2-19) with respect to θ_r and accounting for the number of poles by multiplying by $P/2$ yields the expression for torque

$$T_e = \frac{P}{2}\lambda_m'(i_{as} \cos \theta_r + i_{bs} \sin \theta_r)$$

$$= \frac{P}{2}\lambda_m' i_{qs}^r \tag{6.2-20}$$

The above expression is positive for motor action. The torque and speed may be related as

$$T_e = J\left(\frac{2}{P}\right)\frac{d\omega_r}{dt} + B_m\left(\frac{2}{P}\right)\omega_r + T_L \tag{6.2-21}$$

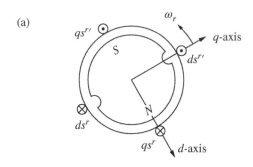

(a)

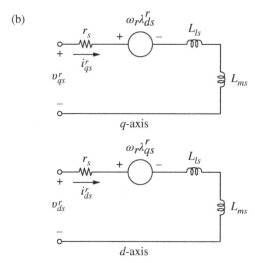

(b)

Figure 6.2-2 (a) The $q's$- and $d's$--axis equivalent of a two-phase permanent-magnet ac machine. (b) Quadrature- and direct-axis equivalent circuits.

where J is in $\mathrm{kg\,m^2}$; it is the inertia of the rotor and any rigidly connected mechanical load. We will be concerned primarily with motor action; therefore, the load torque T_L is assumed positive in the direction indicated in Figure 6.2-1. The constant B_m is a damping coefficient associated with the rotational system of the machine and mechanical load. It has the units N m s/rad and is small and often neglected.

6.3   Instantaneous and Steady-State Phasors

We have established that

$$\tilde{f}_{as}(t) = f^r_{qs}(t) - j f^r_{ds}(t) \tag{6.3-1}$$

In the case of the permanent-magnet ac machine used as the motor in a brushless dc drive, the electrical angular velocity (frequency) of the stator applied voltages is controlled by the inverter to be the same as the electrical angular velocity of the rotor; in other words ω_e is controlled to always be ω_r. Therefore, (6.3-1) will apply as long as $\omega_e = \omega_r$. We find that this is an important observation since the instantaneous and steady-state phasors may be used to portray transient and steady-state machine and drive operation on a phasor diagram if the mode of operation can be expressed in the synchronous reference frame and if steady-state operation yields constant synchronous reference frame variables. The permanent-magnet ac machine, which is an unsymmetrical machine due to the rotor, yields constant synchronous reference frame variables in the steady state only if the rotor speed is synchronous speed. This is not the case in general; however, we find that when the permanent-magnet ac machine is controlled as a brushless dc drive, the fundamental frequency of the applied stator voltages is controlled nearly, instantaneously to be equal to the rotor speed. Therefore, ideally all modes of operation are essentially in the synchronous reference frame, and the phasor representation is valid if we neglect the harmonics introduced in the stator voltages due to the switching of the drive inverter.

Substituting (6.2-15) and (6.2-16) into (6.3-1) yields

$$(v_{qs}^r - jv_{ds}^r) = r_s(i_{qs}^r - ji_{ds}^r) + L_{ss}(pi_{qs}^r - jpi_{ds}^r)$$
$$+ \omega_r L_{ss}(i_{ds}^r + ji_{qs}^r) + \omega_r \lambda_m^{\prime r} \qquad (6.3\text{-}2)$$

which may be written as

$$\tilde{v}_{as} = (r_s + j\omega_r L_{ss})\tilde{i}_{as} + \tilde{e}_a + L_{ss}p\tilde{i}_{as} \qquad (6.3\text{-}3)$$

where the functional notation has been dropped, and we are assuming that

$$\tilde{e}_a = \omega_r \lambda_m^{\prime r}\underline{/0^\circ} \qquad (6.3\text{-}4)$$

You should be wondering why all of the assumptions; first, we have assumed that ω_e and ω_r are always equal and we have used them interchangeably, and now we are assuming that $\tilde{e}_a$ is at zero degrees. We are setting the stage for the analysis of the brushless dc drive where these assumptions become real by the action of the drive inverter.

Now, for the balanced steady-state operation, the last term of (6.3-3) becomes zero, and the steady-state phasor voltage equation becomes

$$\tilde{V}_{as} = (r_s + j\omega_r L_{ss})\tilde{I}_{as} + \tilde{E}_a \qquad (6.3\text{-}5)$$

where

$$\tilde{E}_a = \frac{\omega_r \lambda_m^{\prime r}}{\sqrt{2}}\underline{/0^\circ} \qquad (6.3\text{-}6)$$

For a three-phase machine we replace L_{ms}, which is part of L_{ss}, with $\frac{3}{2}L_{ms}$ in (6.3-3) and (6.3-5).

A convenient expression for the calculation of steady-state torque for balanced operation using phase variables can be obtained by expressing the instantaneous phase currents as

$$I_{as} = \sqrt{2}I_s \cos[\omega_e t + \theta_{esi}(0)] \tag{6.3-7}$$

$$I_{bs} = \sqrt{2}I_s \sin[\omega_e t + \theta_{esi}(0)] \tag{6.3-8}$$

Substituting (6.3-7) and (6.3-8) into (6.2-20), and since $\omega_e = \omega_r$, the steady-state torque may be expressed as

$$
\begin{aligned}
T_e &= \frac{P}{2} \lambda''_m I^r_{qs} \\
&= \frac{P}{2} \lambda''_m \sqrt{2}I_s \cos[\theta_{esi}(0) - \theta_r(0)]
\end{aligned}
\tag{6.3-9}
$$

The phasor diagram can be used to advantage in portraying the steady-state operation of the permanent-magnet ac machine. During balanced steady-state operation, and with $\omega_e = \omega_r$, the rotor and phasor diagrams are rotating in unison; therefore, they can be superimposed with $\tilde{E}_a$ at zero degrees. Recall that $\tilde{E}_a$ and the q-axis are controlled to be at zero degrees. Moreover, since mmf$_r$ is also rotating at ω_r, N^r and S^r can be superimposed with the rotor and the phasor diagrams. Now, since all are rotating in unison and the voltage and current phasors are constant we can stop the rotation at any time, or we can run at ω_e and observe the operation of the machine.

In Figure 6.3-1, we have located the phasors associated with (6.3-6) and the stator poles relative to the stator current. We have assumed that the fundamental component of the north pole of the rotor is fixed in the d-axis. We have motor action, and

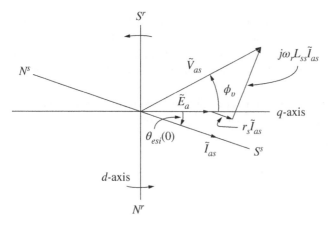

Figure 6.3-1 Phasor diagram showing voltages, currents, and rotor and stator poles of a permanent-magnetic ac machine with $\omega_r = \omega_e$ and $\tilde{E}_a$ at zero degrees.

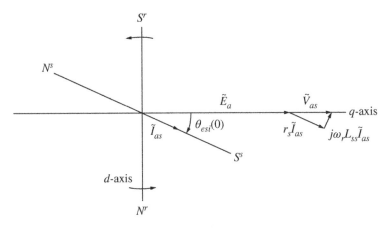

Figure 6.3-2 Phasor diagram for $\tilde{V}_{as}$ and $\tilde{E}_a$ in phase.

we can see this since the poles established by the stators N^s and S^s, which are rotating counterclockwise at $\omega_e(\omega_r)$, are "trying" to pull the rotor poles N^r and S^r, into alignment with them. This "pulling" is being resisted by the load torque. When this "pulling torque" is equal to the load torque on the rotor shaft, steady-state operation occurs with torque in the counterclockwise direction; motor action. Is it clear that complete alignment of the poles occurs only if there is no load torque?

The phasor diagram in Figure 6.3-1 shows $\tilde{V}_{as}$ ahead of $\tilde{E}_a$. This can be controlled by the inverter and changes the torque versus speed characteristics of the inverter–machine combination [1]. The phasor diagram shown in Figure 6.3-2 is for normal operation of a brushless dc drive where $\tilde{V}_{as}$ is controlled to be in phase with $\tilde{E}_a$. In this case, the torque speed characteristics are very similar to a permanent-magnet dc machine, thus the name brushless dc drive.

6.4 Field Orientation of a Brushless dc Drive

The brushless dc drive that we consider consists of the three-phase six-step inverter shown in Figure 6.4-1 and the three-phase permanent-magnet ac machine shown in Figure 6.4-2. The three sensors shown in Figure 6.4-2 are generally Hall-effect devices; when the south pole of the permanent-magnet rotor is under a sensor, its output is nonzero; with the rotor north pole under the sensor, its output is zero. The states of the sensors are used to determine the switching logic for the inverter, which, in turn, determines the output frequency of the inverter. In the actual machine, the sensors are not positioned over the rotor as

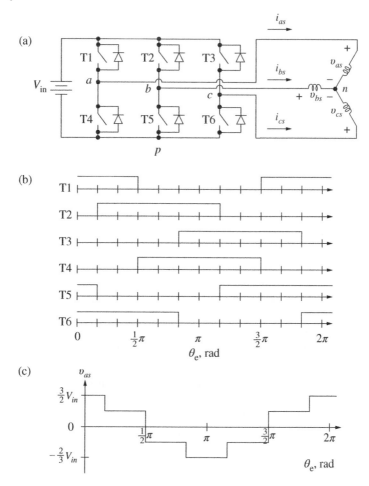

Figure 6.4-1 Inverter–machine drive. (*a*) Inverter configuration, (*b*) transistor switching logic, and (*c*) plot of v_{as}.

shown in Figure 6.4-2. Instead, they are placed over a ring that is mounted on the shaft external to the stator and magnetized the same as the rotor.

The inverter shown in Figure 6.4-1a consists of six transistors, each with an antiparallel diode supplied from a dc-source, V_{in}. The logic (switching) signals for the transistors are shown in Figure 6.4-1b. Each phase is either connected to the top or bottom rail; this is referred to as continuous current or 180° mode of operation. The position of the stator mmf can be advanced or retarded with respect to the rotor mmf by altering the firing of the inverter. This changes the torque speed characteristics of the brushless dc drive.

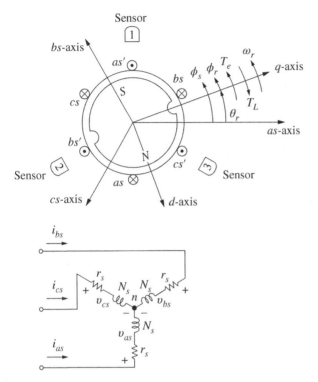

Figure 6.4-2 Two-pole three-phase permanent-magnet ac machine with sensors.

The three-phase voltages of the form given in Figure 6.4-1c, contain a fundamental and 5, 7, 11, and 13 harmonics. Viewed from the stationary reference frame, the fundamental, 7, and 13 produce mmfs that rotate counterclockwise at ω_e, $7\omega_e$, and $13\omega_e$. The 5 and 11 produce mmfs rotating at $5\omega_e$ and $11\omega_e$ in the clockwise direction. In the synchronously rotating reference frame that rotates at ω_e counterclockwise, the fundamental becomes constant, the 5 and 7 harmonics are 6, and the 11 and 13 are 12. With the rotor speed and the position of the rotor poles available, the dc-to-ac inverter is controlled so that the frequency of the fundamental component of the voltages applied to the stator windings is equal instantaneously to the electrical angular velocity of the rotor ω_r. When the torque load on the shaft of the machine is increased the machine slows and the drive inverter control decreases the frequency of the applied stator voltages, which decreases the inductive reactances $(\omega_r L_{ss})$ and $\tilde{E}_a$. Therefore, the decrease in rotor speed allows the current to increase, which in turn increases the strength of the stator rotating magnetic field to accommodate the increase in torque load. Although the primary purpose of the inverter is to control the frequency, it is also used to orient the rotating magnetic field of the stator relative to the permanent magnet of the rotor. This changes the relative position of the stator and rotor poles, which changes the torque characteristics of the machine.

For purposes of the analysis of steady-state operation, it is generally assumed that $\tilde{E}_a$ is at zero degrees, and $\tilde{V}_{as}$, the fundamental component of the as-phase voltage, is at the phase angle $\theta_{esv}(0)$. If, for example, $\phi_v = 0$, which is given by (6.4-1), the inverter control ensures that the positive peak value of the fundamental component of v_{as} and the q-axis are rotating in unison. That is, $\tilde{E}_a$ and $\tilde{V}_{as}$ would be at zero degrees if the phasor diagram is showing the phasors each time the q-axis is horizontal to the right, i.e. coinciding with the as–axis, or if we are running counterclockwise with the q-axis, we would always see the peak value of v_{as}. This is the common mode of operation of a brushless dc drive.

In this section, three control strategies are considered; $\phi_v = 0$, $\phi_v = \phi_{vMT/V}$, and $\phi_v = \phi_{vMT/A}$. When ϕ_v is controlled at $\phi_{vMT/V}$, the applied stator voltages are shifted counterclockwise relative to the q-axis of the rotor to produce the maximum torque per volt that is possible at the instantaneous speed of the rotor (angular frequency of stator voltages). It is found that $\theta_{vMT/V}$ corresponds to the stator impedance angle.

When ϕ_v is controlled at $\phi_{vMT/A}$, the applied stator voltages are shifted in phase relative to the q-axis to produce the maximum torque per ampere possible at the instantaneous speed of the rotor. This occurs when $\tilde{I}_{as}$ is in time phase with the q–axis, or in other words, when $\tilde{I}_{as}$ is orthogonal to the d-axis; thus, the poles of the stator are orthogonal to the permanent magnet.

Before getting further into a discussion of the modes of operation it is helpful to talk a little more about ϕ_v. From Figure 6.3-1

$$\phi_v = \theta_{esv} - \theta_r \tag{6.4-1}$$

For steady-state operation, (6.4-1) becomes

$$\phi_v = \theta_{esv}(0) - \theta_r(0) \tag{6.4-2}$$

Also, for brushless dc machine operation, it is convenient to select $\theta_r(0)$ equal to zero, whereupon

$$\phi_v = \theta_{esv}(0) \tag{6.4-3}$$

Therefore, ϕ_v is the phase of $\tilde{V}_{as}$, and the phase of $\tilde{E}_a$ is zero degrees since it is controlled to coincide with the q-axis.

Brushless dc Motor Operation with $\phi_v = 0$

The parameters of the fractional-horsepower four-pole three-phase permanent-magnet ac machine considered in this section are $r_s = 3.4\,\Omega$, $L_{ls} = 1.1$ mH,

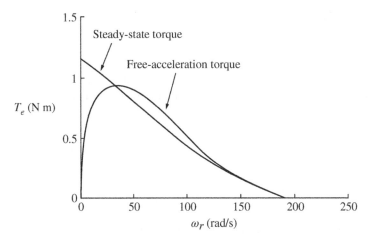

Figure 6.4-3 The free-acceleration torque–speed characteristics for a three-phase machine with $\theta_v = 0$ with the steady-state torque shown for comparison purposes.

$L_{ms} = 11$ mH, and $\lambda_m' = 0.0826$ V s/rad. Recall that L_{ms} must be multiplied by $\frac{3}{2}$ for a three-phase machine due to the mutual coupling of the stator windings when applying the transformation. For brushless dc drive operation, ω_e is made equal to ω_r, and $\phi_v = 0$. The fundamental component of the as-phase voltage phasor is controlled to be "in phase" with the q-axis. Thus,

$$V_{as} = \sqrt{2}V_s \cos\omega_r t \tag{6.4-4}$$

with a balanced abc-sequence, and $V_s = 11.25$ V. With sinusoidal applied voltages, and $\phi_v = 0$, the maximum positive value of $\tilde{V}_{as}$ coincides with the q-axis, and this unison is fixed by controlling the inverter.

The free-acceleration torque–speed characteristics (starting from stall with $T_L = 0$) of the brushless dc drive with $\phi_v = 0$ are shown in Figure 6.4-3. The total inertia J of the rotor and mechanical load is 5×10^{-4} kg m^2, and the damping coefficient B_m is neglected. The steady-state torque is also plotted in Figure 6.4-3 for comparison purposes. The negative slope of the torque–speed characteristics ensures stable operation for motor operation. That is, for a given load torque, a slight slowing of the rotor from an operating point will cause T_e to increase, forcing the rotor back to the operating speed, where $T_L = T_e$. A small increase in rotor speed causes T_e to decrease, whereupon the load torque slows the rotor back to the original operating point, a stable operating condition.

Since $\phi_v = 0$ and $V_{ds}^r = 0$, solving (6.2-18) for I_{ds}^r yields

$$I_{ds}^r = \frac{\omega_r L_{ss}}{r_s}I_{qs}^r \tag{6.4-5}$$

Substituting (6.4-5) into (6.2-17) gives

$$V_{qs}^r = \frac{r_s^2 + \omega_r^2 L_{ss}^2}{r_s} I_{qs}^r + \omega_r \lambda_m''^r \tag{6.4-6}$$

During steady-state operation, all quantities in (6.4-5) and (6.4-6) are constants. If we choose to work with the *abc* variables, the phasor voltage equation is given by (6.4-7), that is

$$\tilde{V}_{as} = (r_s + j\omega_r L_{ss})\tilde{I}_{as} + \tilde{E}_a \tag{6.4-7}$$

where from (6.3-6)

$$\tilde{E}_a = \frac{\omega_r \lambda_m''^r}{\sqrt{2}} /0^\circ \tag{6.4-8}$$

For the three-phase brushless dc machine, the torque is given as

$$\begin{aligned} T_e &= \frac{3}{2}\frac{P}{2}\lambda_m''^r i_{qs}^r \\ &= \frac{3}{2}\frac{P}{2}\lambda_m''^r \sqrt{2}I_s \cos[\theta_{esi}(0) - \theta_r(0)] \end{aligned} \tag{6.4-9}$$

where $\theta_r(0) = 0$.

In summary, the brushless dc drive is so named because the torque-versus-speed characteristics resemble those of a permanent-magnet dc motor. Therefore, it seems logical that there is a reference frame where the voltage and current are dc rather than ac. Since the rotor and synchronous reference frames are the same, these variables are dc in the steady state, and the steady-state torque–speed characteristics (Figure 6.4-3) resemble those of a dc motor. It is also important to mention that the load torque determines the rotor speed, ω_r, whereupon the converter makes the frequency of the applied stator voltages equal to ω_r. For this reason, ω_r is used in (6.4-7) and (6.4-8) to call attention to this fact.

The phasor diagram for operation at 50π rad/s with $\phi_v = 0$ is shown in Figure 6.4-4.

We see from Figure 6.4-4 that the rotor poles can be considered as being "pulled" in the counterclockwise direction by the poles created by the stator currents; motor action.

Maximum Torque per Volt Operation of a Brushless dc Drive ($\phi_v = \phi_{vMT/v}$)

Although $\phi_v = 0$ is a common mode of operation of the brushless dc drive, researchers in [2, 3] discovered that advancing ϕ_v with respect to the *q*-axis could increase the torque at rotor speeds greater than zero. This was shown analytically in [4] and illustrated by simulating the phase shifting (increasing ϕ_v) of the

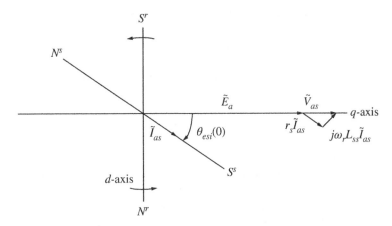

Figure 6.4-4 Phasor diagram for brushless dc drive operation at $\omega_r = 50\pi$ rad/s with $\phi_v = 0$.

applied voltages to obtain maximum torque per volt ($\phi_v = \phi_{vMT/V}$) at a given speed.

If the applied voltages and thus the stator poles are shifted relative to the magnetic field established by the permanent-magnet rotor, the torque versus speed characteristics can be changed over a wide range by shifting ϕ_v from zero to 2π [4]. Here, we limit our discussion to shifting ϕ_v for the purpose of maximizing torque during motor operation.

Torque is proportional to i_{qs}^r, and when ϕ_v is shifted from zero, v_{ds}^r is nonzero. For the purpose of deriving an expression for the maximum torque per volt at a given rotor speed ($\phi_{vMT/V}$), we start with (6.4-10) and (6.4-11) for V_{qs}^r and V_{ds}^r, respectively. In particular, for steady-state operation

$$V_{qs}^r = r_s I_{qs}^r + \omega_r L_{ss} I_{ds}^r + \omega_r \lambda_m^{\prime r} \tag{6.4-10}$$

$$V_{ds}^r = r_s I_{ds}^r - \omega_r L_{ss} I_{qs}^r \tag{6.4-11}$$

Please recall that these equations are valid for two- or three-phase devices for balanced steady-state operation if, for the three-phase device, $\frac{3}{2}L_{ms}$ is used instead of L_{ms}. We also need the expressions for V_{qs}^r and V_{ds}^r as functions of ϕ_v; from (6.4-1) through (6.4-3), which are valid for transient and steady-state operations

$$V_{qs}^r = \sqrt{2}V_s \cos\phi_v \tag{6.4-12}$$

$$V_{ds}^r = -\sqrt{2}V_s \sin\phi_v \tag{6.4-13}$$

Since $\theta_r(0) = 0$, ϕ_v is the phase of $\tilde{V}_{as}$.

Solving (6.4-11) for I_{ds}^r, and substituting the result into (6.4-10) yields

$$V_{qs}^r = \frac{r_s^2 + \omega_r^2 L_{ss}^2}{r_s} I_{qs}^r + \frac{\omega_r L_{ss}}{r_s} V_{ds}^r + \omega_r \lambda_m^{\prime r} \qquad (6.4\text{-}14)$$

Now, solving (6.4-14) for I_{qs}^r, and substituting (6.4-12) and (6.4-13) for V_{qs}^r and V_{ds}^r, respectively, with $\theta_r(0) = 0$, we have

$$I_{qs}^r = \frac{r_s}{r_s^2 + \omega_r^2 L_{ss}^2} \left(\sqrt{2} V_s \cos \phi_v + \frac{\omega_r L_{ss}}{r_s} \sqrt{2} V_s \sin \phi_v - \omega_r \lambda_m^{\prime r} \right) \qquad (6.4\text{-}15)$$

It is interesting to note from (6.4-15) that V_{ds}^r aids V_{qs}^r to increase I_{qs}^r for a given rotor speed. Since this results in a negative I_{ds}^r, it is often referred to as field weakening even though $\lambda_m^{\prime r}$ is not decreased in magnitude.

Since T_e is proportional to I_{qs}^r, (6.4-9), we can obtain the maximum torque for a given rotor speed by taking the derivative of I_{qs}^r with respect to ϕ_v and setting the result equal to zero and then solving for ϕ_v. Thus, from (6.4-15)

$$0 = -\sin \phi_v + \frac{\omega_r L_{ss}}{r_s} \cos \phi_v \qquad (6.4\text{-}16)$$

whereupon

$$\frac{\sin \phi_v}{\cos \phi_v} = \frac{\omega_r L_{ss}}{r_s} \qquad (6.4\text{-}17)$$

or

$$\phi_{vMT/V} = \tan^{-1} \frac{\omega_r L_{ss}}{r_s} \qquad (6.4\text{-}18)$$

Equation (6.4-18) tells us that for a given positive rotor speed, $\phi_{vMT/V}$ will yield maximum possible torque per volt at that rotor speed.

The free-acceleration torque–speed characteristics for $\phi_{vMT/V}$ are shown in Figure 6.4-5. These characteristics may be compared to Figure 6.4-3, where $\phi_v = 0$. Note the extended speed range with $\phi_{vMT/V}$ (Figure 6.4-5) compared with $\phi_v = 0$ (Figure 6.4-3). Also, note that i_{ds}^r is small positive in Figure 6.4-4, but a larger negative value in the phasor diagram is shown in Figure 6.4-6. In other words, an increase in torque (I_{ds}^r) and speed range occurs due to a decrease in I_{ds}^r.

Maximum Torque per Ampere Operation of a Brushless dc Drive ($\phi_v = \phi_{vMT/A}$)

Maximum torque per ampere operation occurs when I_{ds}^r (imaginary part of $\tilde{I}_{as}$) is made zero by controlling the position of $\tilde{V}_{as}$ relative to the permanent magnet of the rotor ($\phi_v = \phi_{vMT/A}$). The torque is directly related to the q-axis current (real part of $\tilde{I}_{as}$). The d-axis current does contribute to the torque but decreases the efficiency of the machine.

To derive an expression for $\phi_{vMT/A}$ for steady-state operation, we substitute (6.4-12) and (6.4-13) into (6.4-10) and (6.4-11) for V_{qs}^r and V_{ds}^r, respectively, and

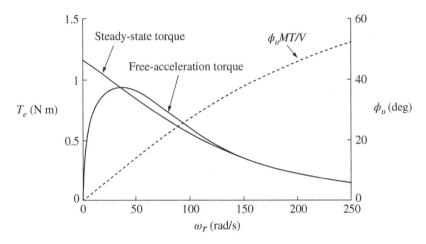

Figure 6.4-5 Torque–speed characteristics of a three-phase machine for free acceleration for $\phi_{vMT/V}$ with the steady-state torque also shown.

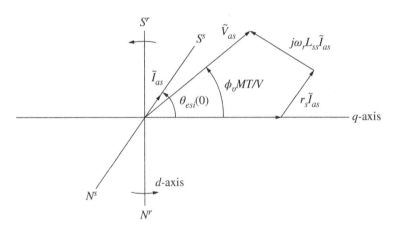

Figure 6.4-6 Phasor diagram for brushless dc drive operation at $\omega_r = 50\pi$ rad/s with $\phi_v = \phi_{vMT/V}$.

solve for $\cos\phi_v$ and $\sin\phi_v$. If we set $I_{ds}^r = 0$ and perform several mathematical manipulations we can express $\phi_{vMT/A}$, at a given rotor speed, as

$$\phi_{vMT/A} = \tan^{-1}\left[\omega_r\tau_s\left(\frac{-1 \pm \omega_r\tau_v\sqrt{1 + \omega_r^2\tau_v^2(1 - \omega_r^2\tau_v^2)}}{\omega_r^4\tau_s^2\tau_v^2 - 1}\right)\right] \qquad (6.4\text{-}19)$$

where, for compactness

$$\tau_s = \frac{L_{ss}}{r_s} \qquad (6.4\text{-}20)$$

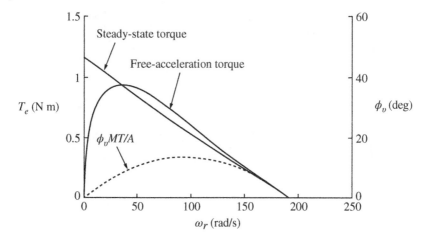

Figure 6.4-7 Torque–speed characteristics for free acceleration with $\phi_{vMT/A}$.

$$\tau_v = \frac{\lambda'^r_m}{\sqrt{2}V_s} \tag{6.4-21}$$

The free-acceleration torque–speed characteristics are shown in Figure 6.4-7 including a plot of $\phi_{vMT/A}$.

The phasor diagram is shown in Figure 6.4-8 for $\omega_r = 50\pi$ rad/s with $\phi_{vMT/A}$. Note that the stator and rotor poles are orthogonal, which yields the maximum torque per ampere for this rotor speed.

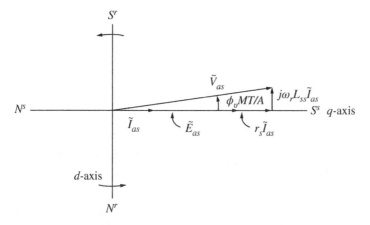

Figure 6.4-8 Phasor diagram for brushless dc drive operation at $\omega_r = 50\pi$ rad/s with $\phi_v = \phi_{vMT/A}$.

6.5 Torque Control of a Brushless dc Drive

The parameters of the four-pole three-phase permanent-magnet ac machine are $r_s = 3.4\,\Omega$, $L_{ls} = 1.1$ mH, $L_{ms} = \frac{3}{2}11$ mH, $\lambda'_m = 0.0826$ V s/rad, and rated $V_s = 11.25$ V. Although we are not going to get into the details of the control, the ideal control that could be used is shown in Figure 6.5-1.

In Figure 6.5-1, the asterisked quantities are commanded values. The torque is controlled with $\phi_v = 0$ at 0.315 N m until the rated voltage is reached at $\omega_r = 120.1$ rad/s. The voltage, V_s, is then maintained at 11.25 V with $\phi_v = 0$, and the

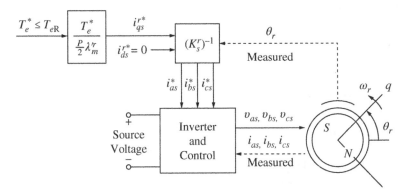

Figure 6.5-1 Block diagram of constant-torque control of a three-phase permanent-magnet ac machine for ω_r less than rated value.

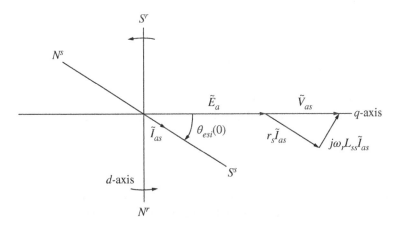

Figure 6.5-2 Phasor diagram for operating Point 1.

load line and operating Point 1 are as shown in Figure 6.5-2. The load line is calculated from

$$T_L = K\omega_r^2 \tag{6.5-1}$$

where $T_L = 0.315$ N m and $\omega_r = 120.1$ rad/s. Substituting into (6.5-1) and solving for K yields

$$K = \frac{0.315}{(120.1)^2} = 2.184 \times 10^{-5} \text{ N m s} \tag{6.5-2}$$

The current $\tilde{I}_{as}$ for operating Point 1 can be determined from (6.4-7). Thus,

$$\tilde{V}_{as} = (r_s + j\omega_r L_{ss})\tilde{I}_{as} + \tilde{E}_a \tag{6.5-3}$$

from which

$$\tilde{I}_{as} = \frac{\tilde{V}_{as} - \tilde{E}_a}{r_s + j\omega_r L_{ss}} \tag{6.5-4}$$

where $\tilde{V}_{as} = 11.25\underline{/0°}$ and

$$\tilde{E}_a = \frac{\omega_r \lambda'^r_m}{r^2}\underline{/0°}$$

$$= \frac{(120.1)(0.0826)}{\sqrt{2}}\underline{/0°} = 7.02\underline{/0°} \tag{6.5-5}$$

$$r_s + j\omega_r L_{ss} = 3.4 + j(120.1)(17.6 \times 10^{-3})$$

$$= 3.4 + j2.113 = 4.0\underline{/-31.8°} \tag{6.5-6}$$

Substituting in (6.5-4) yields

$$\tilde{I}_{as} = \frac{11.25\underline{/0°} = 7.02\underline{/0°}}{4.0\underline{/31.8°}} = 1.058\underline{/-31.8°}\text{ Ax} \tag{6.5-7}$$

The phasor diagram is shown in Figure 6.5-2.

The commanded torque T_e^* is suddenly reduced to one half the original. Thus,

$$T_e^* = \left(\frac{1}{2}\right)(0.315)$$

$$= 0.1575 \text{ N m} \tag{6.5-8}$$

The trajectory from operating Point 1 to operating Point 2 is shown in Figure 6.5-3. The rotor speed at operating Point 2 can be calculated as

$$T_L = K\omega_r^2 \tag{6.5-9}$$

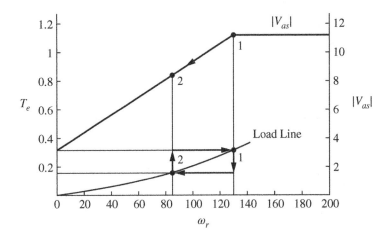

Figure 6.5-3 Controlling T_e with $\phi_v = 0$.

where $T_L = 0.1575$ Nm and $K = 2.184 \times 10^{-5}$ Nm s^2. This yields

$$\omega_r = \left(\frac{0.1575}{2.184 \times 10^{-5}} \right)^{\frac{1}{2}} = 84.9 \text{ rad/s} \qquad (6.5\text{-}10)$$

Now, from (6.4-9)

$$T_e = \left(\frac{3}{2} \right) \left(\frac{P}{2} \right) \lambda''_m \sqrt{2} I_s \cos \theta_{esi}(0) \qquad (6.5\text{-}11)$$

We can write

$$\sqrt{2} I_s \cos \theta_{esi}(0) = \frac{T_e}{\left(\frac{3}{2} \right) \left(\frac{P}{2} \right) \lambda''_m} = \frac{0.1575}{\left(\frac{3}{2} \right) \left(\frac{4}{2} \right) 0.0826} = 0.636 \text{ A} \qquad (6.5\text{-}12)$$

Since $\phi_v = 0$, $\tilde{V}_{as}$ and $\tilde{E}_a$ are at zero degrees; therefore, the phase angle of $\tilde{I}_{as}$ is

$$\theta_{esi}(0) = -\tan^{-1} \frac{\omega_r L_{ss}}{r_s}$$

$$= -\tan^{-1} \frac{(8.49) \left(1.1 + \frac{3}{2} 11 \right) \times 10^{-3}}{r_s}$$

$$= -\tan^{-1} 0.439 = -23.7^{\circ} \qquad (6.5\text{-}13)$$

From (6.5-12)

$$I_s = \frac{0.636}{\sqrt{2} \cos 23.7} = 0.4912 \qquad (6.5\text{-}14)$$

Therefore,

$$\tilde{I}_{as} = 0.4912 \underline{/-23.7^{\circ}} \qquad (6.5\text{-}15)$$

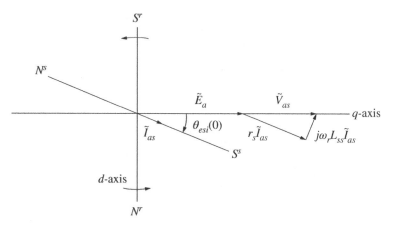

Figure 6.5-4 Phasor diagram for operating Point 2.

Substituting into (6.4-7) and (6.4-8) with L_{ms} multiplied by $\frac{3}{2}$ yields

$$\tilde{V}_{as} = (r_s + j\omega_r L_{ss})\tilde{I}_{as} + \frac{1}{\sqrt{2}}\omega_r \lambda''_m \underline{/0^\circ}$$

$$= \left[3.4 + (84.9)\left(1.1 + \frac{3}{2}11 \times 10^{-3}\right)\right](0.4912\underline{/-23.7^\circ}) + \frac{(84.9)(0.0826)}{\sqrt{2}}\underline{/0^\circ}$$

$$= (3.714\underline{/23.7^\circ})(0.4912\underline{/-23.7^\circ}) + 4.96\underline{/0^\circ} = 6.78\underline{/0^\circ} \qquad (6.5\text{-}16)$$

The phasor diagram for operating Point 2 is shown in Figure 6.5-4.

After steady-state operation is reached at operating Point 2, the commanded torque is increased back to 0.315 N m. The trajectory from operating Point 2 to operating Point 1 is shown in Figure 6.5-3. The mechanical dynamics are determined by

$$T_e^* = J\frac{d\omega_r}{dt} + B_m\omega_r + T_L \qquad (6.5\text{-}17)$$

where $J = 5 \times 10^{-4}$ kg m^2 and $B_m = 0$.

6.6 Closing Comments

We have considered a common drive system by focusing on the operation of the machine and assuming that the control is operating perfectly. This approach allows us to emphasize the importance of reference frame theory in establishing the necessary conditions and constraints to implement the correct control strategy. Our interest has been to portray the necessary operation of the control without becoming engrossed in the design of the control which is done in [1].

References

1 Krause, P., Wasynczuk, O., Sudhoff, S., and Pekarek, S. (2013). *Analysis of Electric Machinery and Drive Systems*, 3e. Wiley, IEEE Press.

2 Nehl, T.W., Fouad, F.A., and Demerdash, N.A. (1981). Digital simulation of power conditioner-machine interaction for electronically commutated DC permanent magnet machines. *IEEE Trans. Magnetics* 17: 3284–3286.

3 Jahns, T.M. (1983). Torque production in permanent-magnet synchronous motor drives with rectangular current excitations. IAS Conf. Rec. (October 1983).

4 Krause, P.C. (1986). *Analysis of Electric Machinery*. New York: McGraw-Hill Book Company.

7

Field Orientation of Induction Machine Drives

7.1 Introduction

Over the past 35 years, the field-oriented induction motor drive has become popular. In this chapter we discuss this drive assuming that the control and power electronics are functioning perfectly, and we neglect harmonics due to the switching of the inverter. Our focus will be on the operation of the machine. We introduced the symmetrical machine in Chapter 4. We refer to that work in this chapter.

7.2 Field Orientation of a Symmetrical Machine

The goal of field orientation of a symmetrical machine is fast torque response, which is accomplished with a controlled drive inverter [1–3]. Unfortunately, field orientation-based control is rather involved since not only is a torque control involved, but it is also necessary to control the stator variables to orient the maximum value of the rotor field, mmf_r^e, in the q-axis, orthogonal to the d-axis component of the stator field, mmf_s^e. To do all this we have two things we can control, the stator applied voltages and the slip frequency. Although the principle of field orientation is quite straightforward, the implementation of the control is not. In particular, the voltage equations in the synchronously rotating reference frame are used to determine the necessary i_{qs}^e and i_{ds}^e to ensure field orientation, and these values are then transformed to the stationary reference frame to determine the stator currents needed for field orientation. This is achieved by controlling the output voltage of the drive inverter to shape the waveforms of the actual stator currents to those determined necessary to provide field orientation. In this section, we will not become involved with the details of implementing this control; instead, we will assume that the control is functioning perfectly and focus on the steady-state performance of an induction machine with field orientation.

Reference Frame Theory: Development and Applications, First Edition. Paul C. Krause.
© 2021 John Wiley & Sons, Inc. Published 2021 by John Wiley & Sons, Inc.

In other words, in this section, our focus will be on what the control does, not how it does it.

The voltage and flux linkage equations of a three-phase single-fed symmetrical machine in the synchronously rotating reference frame for balanced operation (v_{0s} and v'_{0r} both are equal zero) are

$$v^e_{qs} = r_s i^e_{qs} + \omega_e \lambda^e_{ds} + p\lambda^e_{qs} \tag{7.2-1}$$

$$v^e_{ds} = r_s i^e_{ds} - \omega_e \lambda^e_{qs} + p\lambda^e_{ds} \tag{7.2-2}$$

$$0 = r'_r i'^e_{qr} + (\omega_e - \omega_r)\lambda'^e_{dr} + p\lambda'^e_{qr} \tag{7.2-3}$$

$$0 = r'_r i'^e_{dr} - (\omega_e - \omega_r)\lambda'^e_{qr} + p\lambda'^e_{dr} \tag{7.2-4}$$

where $L_{Ms} = \frac{3}{2}L_{ms}$ and

$$\lambda^e_{qs} = L_{ss} i^e_{qs} + L_{Ms} i'^e_{qr} \tag{7.2-5}$$

$$\lambda^e_{ds} = L_{ss} i^e_{ds} + L_{Ms} i'^e_{dr} \tag{7.2-6}$$

$$\lambda'^e_{qr} = L'_{rr} i'^e_{qr} + L_{Ms} i^e_{qs} \tag{7.2-7}$$

$$\lambda'^e_{dr} = L'_{rr} i'^e_{dr} + L_{Ms} i^e_{ds} \tag{7.2-8}$$

where $L_{ss} = L_{ls} + L_{Ms}$ and $L'_{rr} = L'_{lr} + L_{Ms}$ in the above and following equations.

The torque may be expressed as

$$T_e = \frac{3}{2}\frac{P}{2}(\lambda'^e_{qr} i'^e_{dr} - \lambda'^e_{dr} i'^e_{qr}) \tag{7.2-9}$$

The aim is to select the applied stator voltages so that $\lambda'^e_{qr} = 0$, whereupon (7.2-9) becomes

$$T_e = -\frac{3}{2}\frac{P}{2}\lambda'^e_{dr} i'^e_{qr} \tag{7.2-10}$$

Please be careful here. Making λ'^e_{qr} zero does not mean that i'^e_{qr} will also be zero; instead, from (7.2-7), we see that with $\lambda'^e_{qr} = 0$,

$$i'^e_{qr} = -\frac{L_{Ms}}{L'_{rr}} i^e_{qs} \tag{7.2-11}$$

Thus, if i'^e_{qr} is (7.2-11), then $\lambda^e_{qr} = 0$ and T_e is (7.2-10).

From (7.2-10) we see that if λ'^e_{dr} is constant, then torque is proportional to i^e_{qs} from (7.2-11). We will work toward that goal, but first let us control i'^e_{dr} to zero. Why? Well, if i'^e_{dr} is zero, the rotor poles will be positioned completely in the q-axis.

Now again, be careful; controlling i'^e_{dr} to zero does not mean that λ'^e_{dr} is zero; from (7.2-8) with $i'^e_{dr} = 0$,

$$\lambda'^e_{dr} = L_{Ms} i^e_{ds} \tag{7.2-12}$$

which will be constant if i^e_{ds} is held constant. If (7.2-11) and (7.2-12) are substituted into (7.2-10), the torque may be expressed as

$$T_e = \frac{3}{2}\left(\frac{P}{2}\right)\frac{L_{Ms}^2}{L'_{rr}} i^e_{qs} i^e_{ds} \tag{7.2-13}$$

Note that we have T_e in terms of stator-related currents, and if i^e_{ds} is held constant, T_e is directly proportional to i^e_{qs}. This equation is used to control i^e_{qs} when T_e is commanded with i^e_{ds} held constant, generally at its rated value.

We have eliminated (7.2-4), and since λ'^e_{qr} is zero, so must be $p\lambda'^e_{qr}$, and if (7.2-11) is substituted for i'^e_{qr} and (7.2-12) for λ'^e_{dr} into (7.2-3), the angular velocity of the slip may be calculated as

$$(\omega_e - \omega_r)_{\text{calc}} = \frac{r'_r}{L'_{rr}}\frac{i^e_{qs}}{i^e_{ds}} \tag{7.2-14}$$

which we also denote as $(\omega_s)_{\text{calc}}$ for compactness. This equation is used to control slip frequency. Note that (7.2-14), like (7.2-13), is in terms of stator-related currents. We have set forth the basic relationships for field orientation control which in effect positions the rotor poles along the q-axis (7.2-11), orthogonal with the d-axis (I^e_{ds}) or at $\theta_{esi}(0)$ with I^e_{qs} and I^e_{ds} or $\tilde{I}_{as}$.

Steady-State Operation

Our purpose now is to consider the steady-state performance of a symmetrical machine assuming that the field orientation is functioning properly. Although this idealized approach is an oversimplification of the control challenges involved, it helps to give insight into the basic features of field orientation; however, since the stator currents are commanded (controlled), the electric transients are minimized. Therefore, we find that steady state "ideally" controlled operation and the actual field orientation drive with harmonics neglected are very similar.

Let us take a minute to express the steady-state torque assuming the field orientation is functioning properly. We can express I^e_{qs} and I^e_{ds}, respectively, as

$$I^e_{qs} = \sqrt{2}I_s \cos\theta_{esi}(0) \tag{7.2-15}$$

$$I^e_{ds} = -\sqrt{2}I_s \sin\theta_{esi}(0) \tag{7.2-16}$$

with $\omega = \omega_e$ and $\theta(0) = 0$; substituting (7.2-15) and (7.2-16) into (7.2-13) yields the expression for torque with field orientation as

$$T_e = -\frac{3}{2}\left(\frac{P}{2}\right)\frac{L_{Ms}^2}{L_{rr}'}[\sqrt{2}I_s\cos\theta_{esi}(0)][\sqrt{2}I_s\sin\theta_{esi}(0)]$$

$$= -\frac{3}{2}\left(\frac{P}{2}\right)\frac{L_{Ms}^2}{L_{rr}'}I_s^2\sin 2\theta_{esi}(0) \tag{7.2-17}$$

With field orientation and harmonics neglected, the flux linkages in the synchronous reference frame are zero or constants, whereupon ideally, the time rate of change is zero. Therefore, with ideal functioning field orientation, the time rate of change of the currents is zero, and the instantaneous phasor equations become the steady-state phasor equations. The only dynamic feature is the relationship between rotor speed and torque.

We work with the steady-state equations and modify V_{qs}^e and V_{ds}^e to account for the results of field orientation action on the $\overset{e}{qr}$ and $\overset{e}{dr}$ variables given earlier, thereby making it necessary to work only with the stator phasor voltage equations. Substituting $I_{qr}'^e$, in terms of I_{qs}^e from (7.2-11), into (7.2-5) yields

$$\lambda_{qs}^e = \left(\frac{L_{ss}}{L_{Ms}} - \frac{L_{Ms}}{L_{rr}'}\right)L_{Ms}I_{qs}^e \tag{7.2-18}$$

We can now express steady-state V_{qs}^e and V_{ds}^e. From (7.2-1) and (7.2-2) with $p\lambda = 0$, λ_{ds}^e from (7.2-6), with $I_{dr}'^e$ zero, and (7.2-18) for λ_{qs}^e, we have

$$V_{qs}^e = r_s I_{qs}^e + \omega_e L_{ss} I_{ds}^e \tag{7.2-19}$$

$$V_{ds}^e = r_s I_{ds}^e - \omega_e\left(\frac{L_{ss}}{L_{Ms}} - \frac{L_{Ms}}{L_{rr}'}\right)L_{Ms}I_{qs}^e \tag{7.2-20}$$

Let us take a minute to express $\tilde{V}_{as}$ for steady-state operation. Substituting (7.2-19) and (7.2-20) into (4.5-4) yields $\sqrt{2}\tilde{V}_{as}$ with field orientation, that is

$$\sqrt{2}\tilde{V}_{as} = V_{qs}^e - jV_{ds}^e$$

$$= r_s I_{qs}^e + \omega_e L_{ss} I_{ds}^e - j(r_s I_{ds}^e - \omega_e K_L L_{Ms} I_{qs}^e)$$

$$= r_s(I_{qs}^e - jI_{ds}^e) + \omega_e L_{ss} I_{ds}^e + j\omega_e K_L L_{Ms} I_{qs}^e \tag{7.2-21}$$

where

$$K_L = \frac{L_{ss}}{L_{Ms}} - \frac{L_{Ms}}{L_{rr}'} \tag{7.2-22}$$

If now we add and subtract $\omega_e K_L L_{Ms} I_{ds}^e$ on the right-hand side of (7.2-21), we can express $\tilde{V}_{as}$ as

$$\tilde{V}_{as} = (r_s + j\omega_e K_L L_{Ms})\tilde{I}_{as} + \frac{1}{\sqrt{2}}\omega_e(L_{ss} - K_L L_{Ms})I_{ds}^e \tag{7.2-23}$$

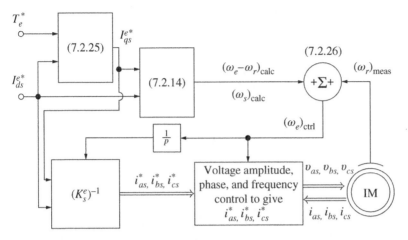

Figure 7.2-1 Block diagram depicting field-oriented control principles.

Substituting (7.2-16) for I_{ds}^e into (7.2-23) yields

$$\tilde{V}_{as} = (r_s + j\omega_e K_L L_{Ms})\tilde{I}_{as} - \omega_e(L_{ss} - K_L L_{Ms})I_s \sin\theta_{esi}(0) \tag{7.2-24}$$

Solving (7.2-13) for I_{qs}^e yields

$$I_{qs}^e = \frac{3}{2}\frac{2}{P}\frac{L_{rr}'}{L_{Ms}^2}\frac{T_e}{I_{ds}^e} \tag{7.2-25}$$

We are also going to use (7.2-14), but let us wait just a minute before we get to that.

Although we are not going to become involved in the details of the field orientation, the block diagram shown in Figure 7.2-1 depicts the basic principles of the control. Therein, the input variables containing an asterisk are commanded variables. That is, if the machine is to operate with rated load torque and rated frequency, then T_e^* would be rated torque and I_{ds}^{e*} would be constant and generally selected to be the value of I_{ds}^e for rated conditions and maintained at that value regardless of T_e^*. Now, (7.2-14), (7.2-25), and (7.2-26), which we get to in a moment, impose field orientation control on the currents, i_{as}^*, i_{bs}^*, and i_{cs}^*. The inverter forms the voltages v_{as}, v_{bs}, and v_{cs} to insure i_{as}, i_{bs}, and i_{cs} and i_{as}^*, i_{bs}^*, and i_{cs}^*, respectively

We have I_{qs}^{e*} and I_{ds}^{e*}, and from (7.2-14), we can calculate $\omega_e - \omega_r$, which we are calling $(\omega_s)_{\text{calc}}$ in Figure 7.2-1 (the slip angular velocity) and/or $(\omega_e - \omega_r)_{\text{calc}}$ since it is calculated from commanded inputs I_{qs}^{e*} and I_{ds}^{e*} and machine parameters. Now, $(\omega_s)_{\text{calc}}$ is the slip necessary for field orientation with the commanded stator-related currents. Now, $(\omega_r)_{\text{meas}}$ is the measured electrical angular velocity of the rotor, and it is added to $(\omega_s)_{\text{calc}}$. Thus, from Figure 7.2-1

$$(\omega_e)_{\text{ctrl}} = (\omega_s)_{\text{calc}} + (\omega_r)_{\text{meas}}$$

$$= (\omega_e - \omega_r)_{\text{calc}} + (\omega_r)_{\text{meas}} \tag{7.2-26}$$

Table 7.2-1 Steady-state voltage and torque equations for field orientation of induction machines.

$$V_{qs}^e = r_s I_{qs}^e + \omega_e L_{ss} I_{ds}^e$$

$$V_{ds}^e = r_s I_{ds}^e - \omega_e K_L L_{Ms} I_{qs}^e$$

$$T_e = \frac{3}{2}\frac{P}{2}\frac{L_{Ms}^2}{L_{rr}'} i_{qs}^e i_{ds}^e$$

$$\tilde{V}_{as} = (r_s + j\omega_e K_L L_{Ms})\tilde{I}_{as}$$

$$\quad - \frac{1}{\sqrt{2}}\omega_e(L_{ss} - K_L L_{Ms})I_{ds}^e$$

$$T_e = -\frac{3}{2}\left(\frac{P}{2}\right)\frac{L_{Ms}^2}{L_{rr}'}[\sqrt{2}I_s\cos\theta_{esi}(0)][\sqrt{2}I_s\sin\theta_{esi}(0)]$$

$$\quad = -\frac{3}{2}\left(\frac{P}{2}\right)\frac{L_{Ms}^2}{L_{rr}'}I_s^2\sin 2\theta_{esi}(0)$$

$$K_L = \frac{L_{ss}}{L_{Ms}} - \frac{L_{Ms}}{L_{rr}'}$$

where $(\omega_e)_{\text{ctrl}}$ is the electrical angular velocity of the controlled stator voltages applied to the machine (v_{as}, v_{bs}, and v_{cs} of Figure 7.2-1) for the rotor magnetic field to be oriented in the q-axis.

There is no direct control of speed; however, if we are operating in the steady state and the torque load increases, $(\omega_r)_{\text{meas}}$ would decrease, which would decrease ω_e until $(\omega_e - \omega_r)_{\text{calc}}$ was established. Again, $(\omega_e - \omega_r)_{\text{calc}}$ is the slip angular velocity, calculated from the commanded T_e^* and I_{ds}^{e*} by (7.2-14). Assuming the parameters remain constant with temperature change, then $(\omega_e - \omega_r)_{\text{calc}}$ will change only if T_e^* and/or I_{ds}^{e*} is changed.

It is important to note that the commanded variables are in the synchronously rotating reference frame (I_{qs}^{e*} and I_{ds}^{e*}). These commanded variables are transformed to the stationary reference frame by $(\mathbf{K}_s^e)^{-1}$ to make up the waveforms of i_{as}, i_{bs}, and i_{cs}. The phase and amplitude of the stator voltages are then controlled by the drive inverter in order to produce the desired stator currents (Table 7.2-1).

7.3 Torque Control of Field-Orientated Symmetrical Machine

The parameters of a three-phase four-pole 3-hp 220-V 60 Hz induction machine are

$$r_s = 0.435\,\Omega \quad L_{Ms} = 69.3\,\text{mH} \quad r_r' = 0.816\,\Omega$$

$$L_{ls} = 2.0\,\text{mH} \qquad\qquad L_{lr}' = 2.0\,\text{mH}$$

The rated speed is 1710 rpm, and the rated torque is 11.9 N m. The load line is

$$T_L = K\omega_r^2 \tag{7.3-1}$$

The machine is initially operating at Point 1 of Figure 7.3-1

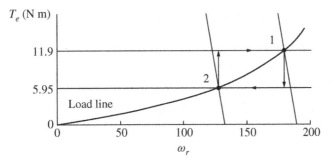

Figure 7.3-1 Operation of induction motor drive with field orientation for step change in T_e^*.

From (7.2-13)

$$T_e = \left(\frac{3}{2}\right)\left(\frac{P}{2}\right)\frac{L_{Ms}^{\ 2}}{L_{rr}'}I_{qs}^e I_{ds}^e \tag{7.3-2}$$

Now,

$$I_{qs}^e = \sqrt{2}\,I_{rated}$$
$$= \sqrt{2}\,5.8 = 8.2\ \text{A} \tag{7.3-3}$$

The rated torque is 11.9 N m, and I_{ds}^e may be determined from (7.3-2) as

$$I_{ds}^e = 11.9\left(\frac{2}{3}\right)\left(\frac{2}{4}\right)\frac{71.3\times10^{-3}}{(69.3\times10^{-3})^2}\frac{1}{8.2}$$
$$= 7.18\ \text{A} \tag{7.3-4}$$

$$\tilde{I}_{as} = \frac{1}{\sqrt{2}}(I_{qs}^e - jI_{ds}^e)$$
$$= \frac{1}{\sqrt{2}}(8.2 - j7.18) = 5.8 - j5.1$$
$$= 7.7\underline{/-41.3}\ \text{A} \tag{7.3-5}$$

From (7.2-22)

$$K_L = \frac{L_{ss}}{L_{Ms}} - \frac{L_{Ms}}{L_{rr}'}$$
$$= \frac{71.3\times10^3}{69.3\times10^3} - \frac{69.3\times10^3}{71.3\times10^{-3}} = 0.057 \tag{7.3-6}$$

From (7.2-14)

$$(\omega_e - \omega_r)_{calc} = \frac{r_r'}{L_{rr}'}\frac{I_{qs}^e}{I_{ds}^e}$$

$$= \frac{0.816}{71.3 \times 10^{-3}} \frac{8.2}{7.18} = 13.08 \,\text{rad/s} \tag{7.3-7}$$

The rated speed = 1710 for the four-pole machine, which is 178.9 rad/s × 2 = 357.96 + 13.08 = 371 rad/s for the two-pole equivalent.

To calculate $\tilde{V}_{as}$ for operating Point 1, we use (7.2-23)

$$\tilde{V}_{as} = (r_s + j\omega_e K_L L_{Ms})\tilde{I}_{as} + \frac{1}{\sqrt{2}} \omega_e (L_{ss} - K_L L_{Ms})I_{ds}^e$$

$$= [0.435 + j(371)(0.057)(69.3 \times 10^{-3})]7.7\underline{/-41.3^{\circ}}$$

$$+ \frac{1}{\sqrt{2}}371[71.3 \times 10^{-3} - (0.057)(69.3 \times 10^{-3})]7.18$$

$$= 10.0 + j6.29 + 126.7 = 136.7 + j6.29$$

$$= \underline{136.8/2.84^{\circ}} \tag{7.3-8}$$

The phasor diagram for operating Point 1 is given in Figure 7.3-2. Please note that the voltage $\tilde{V}_{as}$ calculated in (7.3-8) is aboverated by 10 V. We have assumed a magnetically linear machine, and an overvoltage could cause saturation. If so, I_{ds}^e could be reduced, which would cause a decrease in T_e in (7.3-2).

The torque command is reduced to 5.95 N m. The machine slows and finally establishes steady-state operation at operating Point 2 in Figure 7.3-1. The new rotor speed can be determined from (7.3-1). From operating Point 1, K can be calculated. Thus,

$$K = \frac{T_L}{\omega_r^2} = \frac{11.9}{(178.98)^2} = 0.372 \times 10^{-3} \tag{7.3-9}$$

At operating Point 2, $T_e = 5.95$ N m, and

$$\omega_r = \left(\frac{5.95}{0.372 \times 10^{-3}}\right)^{\frac{1}{2}} = 126.5 \,\text{rad/s} \tag{7.3-10}$$

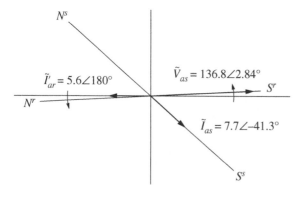

Figure 7.3-2 Phasor diagram for operating Point 1.

Figure 7.3-3 Phasor diagram for operating Point 2.

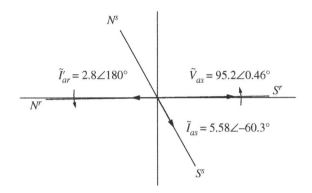

Now, $(\omega_e - \omega_r)$ is $\frac{1}{2}$ of (7.3-7) or 6.5 rad/s and

$$\omega_e = 126.5 \times 2 + 6.5 = 259.4 \, \text{rad/s} \tag{7.3-11}$$

Also, I_{qs}^e from (7.3-2) is 4.1 A, and $I_{ds}^e = 7.18$ A. $\tilde{I}_{as}$ is now $\tilde{I}_{as} = \frac{1}{\sqrt{2}}(4.1 - j7.18) =$ 5.85$\underline{/-60.3^\circ}$ A. This gives $\tilde{V}_{as} = 95.2\underline{/0.46^\circ}$ from (7.3-8). The phasor diagram for operating Point 2 is given in Figure 7.3-3. The trajectory given in Figure 7.3-1 is complete by increasing the commanded torque back to 11.9 N m.

7.4 Closing Comments

This chapter allows us to look at field-oriented control assuming that the control is operating perfectly. Torque control with field orientation is given as an example.

References

1 Krause, P.C., Wasynczuk, O., Sudhoff, S.D., and Pekarek, S.D. (2013). *Analysis of Electric Machinery and Drive Systems*, 3e. New York: Wiley, IEEE Press.

2 Trzynadlowski, A.M. (1994). *The Field Orientation Principle in Control of Induction Motors*. Kluwer Academic Publishers.

3 Blaschke, F. (1974). Das Verfahren der Feldorientierung zur Regelung der Drehfeld-maschine. PhD thesis, TU Braunschweig.

8

Additional Applications of Reference Frame Theory

8.1 Introduction

This chapter sets forth simplified versions of three applications of reference frame theory that the reader may find interesting, in particular, neglecting stator transients, the theory of symmetrical components derived by reference frame theory, and the concept of multiple reference frames. Each topic has been published in IEEE Transactions; however, these papers are in detail and probably of interest only to the person doing research and who wants to dig deeper into the area of electric machine analysis.

8.2 Neglecting Stator Transients

The use of reference frame theory to establish the neglecting of stator transients in any reference frame is set forth in this section. Neglecting stator transients became a concern when simulating Park's equations for each machine in a large power system. It became necessary to neglect the stator transients and the electric transient of the connecting transmission network in order to reduce the number of integrations. For 20 years the literature was filled with trial-and-error approaches to find the appropriate terms to neglect in Park's equations to yield the best results. A paper written in 1979 solved the problem [1].

Since the electric machines and transmission lines are essentially inductive, the voltage equations for a series rL circuit in the synchronous reference frame are

$$v_{qs}^e = r_s i_{qs}^e + \omega_e \lambda_{ds}^e + p\lambda_{qs}^e \tag{8.2-1}$$

$$v_{ds}^e = r_s i_{ds}^e - \omega_e \lambda_{qs}^e + p\lambda_{ds}^e \tag{8.2-2}$$

$$v_{0s} = r_s i_{0s} + L_{ls} p i_{0s} \tag{8.2-3}$$

Reference Frame Theory: Development and Applications, First Edition. Paul C. Krause.
© 2021 John Wiley & Sons, Inc. Published 2021 by John Wiley & Sons, Inc.

This work is valid only for balanced symmetrical conditions; therefore, the zero variables do not exist. Now, in the synchronous reference frame, the $p\lambda^e$ terms contain only the electric transients; however, during unbalanced operation, the q and d variables contain time-varying quantities. Therefore, neglecting the electric transients for balanced conditions is to neglect $p\lambda^e$ terms. The static network representation in essence is neglecting the $p\lambda$ terms in the synchronous reference frame. However, the synchronous machine is simulated by Park's equations which are in the rotor reference frame. So, the question is what terms do you neglect in reference frames other than the synchronous in order to neglect stator transients? This was the problem facing engineers, and it sparked considerable trial-and-error "research."

One can answer the question by determining the terms that must be neglected in the arbitrary reference frame in order to neglect just the $p\lambda^e$ terms. We determine this by transforming the $p\lambda^e$ terms to the arbitrary reference frame while preserving the identity of these terms in the transformation. However, before doing this we must devise the transformation between reference frames.

Let us say, we want to transform from the x to the y reference frames. That is,

$$\mathbf{f}^y_{qd0s} = {}^x\mathbf{K}^y \mathbf{f}^x_{qd0s} \tag{8.2-4}$$

Now,

$$\mathbf{f}^x_{qd0s} = \mathbf{K}^x_s \mathbf{f}_{abcs} \tag{8.2-5}$$

$$\mathbf{f}^y_{qd0s} = \mathbf{K}^y_s \mathbf{f}_{abcs} \tag{8.2-6}$$

Thus,

$$\mathbf{K}^y_s f_{abcs} = {}^x\mathbf{K}^y \, \mathbf{K}^x_s \, \mathbf{f}_{abcs} \tag{8.2-7}$$

and

$${}^x\mathbf{K}^y = \mathbf{K}^y_s (\mathbf{K}^x_s)^{-1} \tag{8.2-8}$$

Thus,

$${}^e\mathbf{K} = \begin{bmatrix} \cos(\theta - \theta_e) & -\sin(\theta - \theta_e) & 0 \\ \sin(\theta - \theta_e) & \cos(\theta - \theta_e) & 0 \\ 0 & 0 & 1 \end{bmatrix} \tag{8.2-9}$$

Please recall that the arbitrary variables carry no superscript. Also, recall that this work is only valid for balanced symmetrical conditions; therefore, the zero variables are not present, and to transform from the synchronously rotating reference frame to the arbitrary reference frame

$${}^e\mathbf{K} = \begin{bmatrix} \cos(\theta - \theta_e) & -\sin(\theta - \theta_e) \\ \sin(\theta - \theta_e) & \cos(\theta - \theta_e) \end{bmatrix} \tag{8.2-10}$$

the inverse is

$$({}^{e}\mathbf{K})^{-1} = \begin{bmatrix} \cos(\theta - \theta_{e}) & \sin(\theta - \theta_{e}) \\ -\sin(\theta - \theta_{e}) & \cos(\theta - \theta_{e}) \end{bmatrix} \tag{8.2-11}$$

Let us now transform (8.2-1) and (8.2-2) to the arbitrary reference frame while preserving the identity of the $p\lambda^{e}$ terms

$$\mathbf{v}_{qds} = {}^{e}\mathbf{K}\,\mathbf{v}^{e}_{qds} \tag{8.2-12}$$

$$\mathbf{v}_{qds} = {}^{e}\mathbf{K}\,\mathbf{r}_{s}({}^{e}\mathbf{K})^{-1}\mathbf{i}_{qds} + {}^{e}\mathbf{K}\,\omega_{e}({}^{e}\mathbf{K})^{-1}\begin{bmatrix} \lambda_{ds} \\ -\lambda_{qs} \end{bmatrix}$$
$$+ {}^{e}\mathbf{K}\,p({}^{e}\mathbf{K})^{-1}\lambda_{qds} \tag{8.2-13}$$

The first two terms on the right-hand side are not related to $p\lambda^{e}_{qd}$, but the third term is. Therefore, the electric transients are neglected in the arbitrary reference frame by neglecting the third term; thus,

$$v_{qs} = r_{s}i_{qs} + \omega_{e}\,\lambda_{ds} \tag{8.2-14}$$

$$v_{ds} = r_{s}i_{ds} - \omega_{e}\,\lambda_{qs} \tag{8.2-15}$$

Recall that Park's equations have $\omega_{r}\,\lambda_{qs}$ and $\omega_{r}\,\lambda_{ds}$, which should be $\omega_{e}\,\lambda^{r}_{ds}$ and $\omega_{e}\,\lambda^{r}_{qs}$ in order to neglect stator transients. In the case of the symmetrical (induction) machine, (8.2-14) and (8.2-15) would replace the stator equations in all reference frames.

In the case of the symmetrical rotor, the voltage equations in the synchronous reference frame are

$$v'^{e}_{qr} = r'_{r}i'^{e}_{qr} + (\omega_{e} - \omega_{r})\lambda'^{e}_{dr} + p\lambda'^{e}_{qr} \tag{8.2-16}$$

$$v'^{e}_{dr} = r'_{r}i'^{e}_{dr} - (\omega_{e} - \omega_{r})\lambda'^{e}_{qr} + p\lambda'^{e}_{dr} \tag{8.2-17}$$

The rotor voltage equations with the rotor transients neglected in the arbitrary reference frame become

$$v'_{qr} = r'_{r}i'_{qr} + (\omega_{e} - \omega_{r})\lambda'_{dr} \tag{8.2-18}$$

$$v'_{dr} = r'_{r}i'_{dr} - (\omega_{e} - \omega_{r})\lambda'_{qr} \tag{8.2-19}$$

8.3 Symmetrical Components Derived by Reference Frame Theory

In [2] we allowed a very general unbalance of a three-phase system, including different frequencies in each phase and time-varying coefficients. Here, we will not be so general; in particular, let us consider the unbalance of the f_{as}, f_{bs}, f_{cs} variables, which can be written as

$$f_{as} = f_{as\alpha} \cos \omega_e t + f_{as\beta} \sin \omega_e t \qquad (8.3\text{-}1)$$

$$f_{bs} = f_{bs\alpha} \cos \omega_e t + f_{bs\beta} \sin \omega_e t \qquad (8.3\text{-}2)$$

$$f_{cs} = f_{cs\alpha} \cos \omega_e t + f_{cs\beta} \sin \omega_e t \qquad (8.3\text{-}3)$$

where the coefficients are constant, and only one frequency exists. This type of unbalance would cover the majority that occurs.

If now we transform (8.3-1) through (8.3-3) to the arbitrary reference frame

$$\mathbf{f}_{qd0s} = \mathbf{K}_s \, \mathbf{f}_{abcs} \qquad (8.3\text{-}4)$$

where

$$\mathbf{f}_{qd0s} = (f_{qs} \, f_{ds} \, f_{0s})^T \qquad (8.3\text{-}5)$$

$$\mathbf{f}_{abcs} = (f_{as} \, f_{bs} \, f_{cs})^T \qquad (8.3\text{-}6)$$

and

$$\mathbf{K}_s = \frac{2}{3} \begin{bmatrix} \cos\theta & \cos\left(\theta - \frac{2}{3}\pi\right) & \cos\left(\theta + \frac{2}{3}\pi\right) \\ \sin\theta & \sin\left(\theta - \frac{2}{3}\pi\right) & \sin\left(\theta + \frac{2}{3}\pi\right) \\ \frac{1}{2} & \frac{1}{2} & \frac{1}{2} \end{bmatrix} \qquad (8.3\text{-}7)$$

$$\begin{aligned} f_{qs} &= f_{qsA} \cos(\omega_e t - \theta) + f_{qsB} \sin(\omega_e t - \theta) \\ &\quad + f_{qsC} \cos(\omega_e t - \theta) + f_{qsD} \sin(\omega_e t - \theta) \end{aligned} \qquad (8.3\text{-}8)$$

$$\begin{aligned} f_{ds} &= f_{dsA} \cos(\omega_e t - \theta) + f_{dsB} \sin(\omega_e t - \theta) \\ &\quad + f_{dsC} \cos(\omega_e t - \theta) + f_{dsD} \sin(\omega_e t - \theta) \end{aligned} \qquad (8.3\text{-}9)$$

$$\begin{aligned} f_{0s} &= \frac{1}{3}(f_{as\alpha} + f_{bs\alpha} + f_{cs\alpha}) \cos \omega_e t \\ &\quad + \frac{1}{3}(f_{as\beta} + f_{bs\beta} + f_{cs\beta}) \sin \omega_e t \end{aligned} \qquad (8.3\text{-}10)$$

where

$$\begin{aligned} f_{qsA} &= \frac{1}{3}\left[f_{as\alpha} - \frac{1}{2}f_{bs\alpha} - \frac{1}{2}f_{cs\alpha} + \frac{\sqrt{3}}{2}(f_{bs\beta} - f_{cs\beta}) \right] \\ &= -f_{dsB} \end{aligned} \qquad (8.3\text{-}11)$$

$$f_{qsB} = \frac{1}{3}\left[f_{as\beta} - \frac{1}{2}f_{bs\beta} - \frac{1}{2}f_{cs\beta} + \frac{\sqrt{3}}{2}(f_{bs\alpha} - f_{cs\alpha}) \right]$$

$$= -f_{dsA} \tag{8.3-12}$$

$$f_{qsC} = \frac{1}{3}\left[f_{as\alpha} - \frac{1}{2}f_{bs\alpha} - \frac{1}{2}f_{cs\alpha} + \frac{\sqrt{3}}{2}(f_{bs\beta} - f_{cs\beta}) \right]$$

$$= -f_{dsD} \tag{8.3-13}$$

$$f_{qsD} = \frac{1}{3}\left[f_{as\beta} - \frac{1}{2}f_{bs\beta} - \frac{1}{2}f_{cs\beta} + \frac{\sqrt{3}}{2}(f_{bs\alpha} - f_{cs\alpha}) \right]$$

$$= -f_{dsC} \tag{8.3-14}$$

We see from (8.3-8) through (8.3-14) that f_{qs} and f_{ds} contain balanced two-phase positive and negative sequences. For example, the first term of (8.3-8) and the second term of (8.3-9) along with the second term of (8.3-8) and the first term of (8.3-9) both produce counterclockwise rotating mmfs; positive sequences. The negative-sequence mmfs are formed by the third term of (8.3-8) and the fourth term of (8.3-9) and the fourth term of (8.3-8) and the third term of (8.3-9).

For steady-state operation, we use the stationary reference frame by setting θ in (8.3-8) and (8.3-9) equal to zero but still preserving the forms of (8.3-8) and (8.3-9). Therefore, using uppercase for steady-state variables, we can write for (8.3-8) and (8.3-9), respectively,

$$\tilde{F}_{qs}^s = \tilde{F}_{qs+}^s + \tilde{F}_{qs-}^s \tag{8.3-15}$$

$$\tilde{F}_{ds}^s = \tilde{F}_{ds+}^s + \tilde{F}_{ds-}^s \tag{8.3-16}$$

Now, if we consider (8.3-8) and (8.3-9) along with (8.3-11) through (8.3-14), we can write

$$\sqrt{2}\tilde{F}_{qs+}^s = F_{qsA} - jF_{qsB} \tag{8.3-17}$$

$$\sqrt{2}\tilde{F}_{ds+}^s = F_{dsA} - jF_{dsB} = j\sqrt{2}\tilde{F}_{qs+}^s \tag{8.3-18}$$

$$\sqrt{2}\tilde{F}_{qs-}^s = F_{qsC} - jF_{qsD} \tag{8.3-19}$$

$$\sqrt{2}\tilde{F}_{ds-}^s = F_{dsC} - jF_{dsD} = -j\sqrt{2}\tilde{F}_{qs-}^s \tag{8.3-20}$$

Substituting (8.3-18) and (8.3-20) into (8.3-16), we can write (8.3-15) and (8.3-16) as

$$\begin{bmatrix} \tilde{F}_{qs}^s \\ \tilde{F}_{ds}^s \end{bmatrix} = \begin{bmatrix} 1 & 1 \\ j^1 & -j^1 \end{bmatrix} \begin{bmatrix} \tilde{F}_{qs+}^s \\ \tilde{F}_{qs-}^s \end{bmatrix} \tag{8.3-21}$$

The zero quantities may be written in phasor form as

$$\sqrt{2}\tilde{F}_{0s} = F_{0s\alpha} - jF_{0s\beta} \tag{8.3-22}$$

where $F_{0s\alpha}$ and $F_{0s\beta}$ are defined by (8.3-10).

Now, the $\tilde{F}_{qs}^s$, $\tilde{F}_{ds}^s$, and $\tilde{F}_{0s}$ are related to $\tilde{F}_{as}$, $\tilde{F}_{bs}$, and $\tilde{F}_{cs}$ by $\mathbf{K}_s^s$, that is

$$[\tilde{F}_{qs}^s \ \tilde{F}_{ds}^s \ \tilde{F}_{0s}]^T = \mathbf{K}_s^s[\tilde{F}_{as} \ \tilde{F}_{bs} \ \tilde{F}_{cs}]^T$$

$$= \frac{2}{3} \begin{bmatrix} 1 & -\dfrac{1}{2} & -\dfrac{1}{2} \\ 0 & -\dfrac{\sqrt{3}}{2} & \dfrac{\sqrt{3}}{2} \\ \dfrac{1}{2} & \dfrac{1}{2} & \dfrac{1}{2} \end{bmatrix} \begin{bmatrix} \tilde{F}_{as} \\ \tilde{F}_{bs} \\ \tilde{F}_{cs} \end{bmatrix} \tag{8.3-23}$$

Also, (8.3-21) is the inverse of the well-known two-phase symmetrical component transformation, which is

$$\begin{bmatrix} \tilde{F}_{qs+}^s \\ \tilde{F}_{qs-}^s \end{bmatrix} = \frac{1}{2} \begin{bmatrix} 1 & -j1 \\ 1 & j1 \end{bmatrix} \begin{bmatrix} \tilde{F}_{qs}^s \\ \tilde{F}_{ds}^s \end{bmatrix} \tag{8.3-24}$$

From (8.3-22) and (8.3-24)

$$[\tilde{F}_{qs+}^s \ \tilde{F}_{qs-}^s \ \tilde{F}_{0s}]^T = \mathbf{S}_{3qd}[\tilde{F}_{qs}^s \ \tilde{F}_{ds}^s \ \tilde{F}_{0s}]^T \tag{8.3-25}$$

where

$$\mathbf{S}_{3qd} = \frac{1}{2} \begin{bmatrix} 1 & -j1 & 0 \\ 1 & j1 & 0 \\ 0 & 0 & 2 \end{bmatrix} \tag{8.3-26}$$

Substituting (8.3-23) into (8.3-25) yields

$$[\tilde{F}_{qs+}^s \ \tilde{F}_{qs-}^s \ \tilde{F}_{0s}] = \mathbf{S}_{3qd}\mathbf{K}_s^s[\tilde{F}_{as} \ \tilde{F}_{bs} \ \tilde{F}_{cs}]^T$$

$$= \mathbf{S}[\tilde{F}_{as} \ \tilde{F}_{bs} \ \tilde{F}_{cs}]^T \tag{8.3-27}$$

where $\mathbf{S}$ is the familiar symmetrical component transformation for a three-phase system, that is

$$\mathbf{S} = \frac{1}{3} \begin{bmatrix} 1 & a & a^2 \\ 1 & a^2 & a \\ 1 & 1 & 1 \end{bmatrix} \tag{8.3-28}$$

where a is complex, denoting a counterclockwise rotation of $\frac{2\pi}{3}$ radians. We have just shown that the arbitrary reference frame contains the complex symmetrical component transformation. Thus, it appears that Tesla's rotating magnetic field contains all real and complex transformations used in power and drive system analyses.

8.4 Multiple Reference Frames

When calculating steady-state unbalanced operation of a symmetrical machine, the method of symmetrical components is generally used. Symmetrical components involved phasors which are complex numbers. There is another method which involves only real numbers [2]. The theory of multiple reference frames allows calculation of steady-state unbalanced operation using different reference frames. This is possible because unbalanced stator variables can be portrayed in the arbitrary reference frame by balanced sets as we saw in (8.3-8) and (8.3-9).

The positive sequence can be solved using real numbers in the synchronously rotating reference frame, $\omega = \omega_e$. The negative sequence can be solved using real numbers in the synchronous rotating at $\omega = -\omega_e$. Therefore, depending on the unbalance, (8.3-8) through (8.3-10) and (8.3-11) through (8.3-14) can be used to solve a general unbalance. In [3], multiple reference frames are used to solve for the currents of an inverter induction motor drive. We need not set forth additional equations since the concept of multiple reference frames is clear from the work in Section 8.3. Moreover, it is covered in [4–7].

8.5 Closing Comments

In this chapter we have considered the concept of three papers. Those interested in the details may refer to the actual papers or the referenced books.

It appears that the expression of Tesla's rotating magnetic contains all known real transformations used in power and drives for the past 100 years. This was set forth in Chapter 3. Although not as direct, it is shown that the most used complex transformation, the symmetrical component transformation, is also contained in Tesla's rotating magnetic field.

References

1 Krause, P.C., Nozari, F., Skvarenina, T.L., and Olive, D.W. (1979). The theory of neglecting stator transients. *IEEE Trans. Power App. Syst.* 98: 141–145.

2 Krause, P.C. (1985). The method of symmetrical components derive by reference frame theory. *IEEE Trans. Power App. Syst.* 104: 1492–1499.

3 Krause, P.C. (1968). Method of multiple reference frames applied to the analysis of symmetrical induction machinery. *IEEE Trans. Power App. Syst.* 87: 227–234.

4 Krause, P.C. (1986). *Analysis of Electric Machinery.* New York: McGraw-Hill Book Company.

5 Krause, P.C., Wasynczuk, O., and Sudhoff, S.D. (1994). *Analysis of Electric Machinery*, 1e. New York: Wiley, IEEE Press.

6 Krause, P.C., Wasynczuk, O., and Sudhoff, S.D. (2002). *Analysis of Electric Machinery and Drive Systems*, 2e. New York: Wiley, IEEE Press.

7 Krause, P.C., Wasynczuk, O., Sudhoff, S.D., and Pekarek, S.D. (2013). *Analysis of Electric Machinery and Drive Systems*, 3e. New York: Wiley, IEEE Press.

Index

Reference Frame Theory: Development and Applications, First Edition. Paul C. Krause.
© 2021 John Wiley & Sons, Inc. Published 2021 by John Wiley & Sons, Inc.

Printed and bound by CPI Group (UK) Ltd, Croydon, CR0 4YY